Applied Sport Mechanics

FOURTH EDITION

Brendan Burkett, BEng, MEng, PhD

School of Health and Sport Sciences
University of the Sunshine Coast, Australia

HUMAN KINETICS

Library of Congress Cataloging-in-Publication Data

Names: Burkett, Brendan, 1963- author.
Title: Applied sport mechanics / Brendan Burkett.
Other titles: Sport mechanics for coaches
Description: Fourth edition. | Champaign, IL : Human Kinetics, [2019] |
 Preceded by Sport mechanics for coaches / Brendan Burkett. 3rd ed. c2010.
 | Includes bibliographical references and index.
Identifiers: LCCN 2017054181 (print) | LCCN 2017055687 (ebook) | ISBN
 9781492558446 (e-book) | ISBN 9781492558439 (print)
Subjects: | MESH: Biomechanical Phenomena | Sports--physiology | Movement |
 Motion
Classification: LCC QP303 (ebook) | LCC QP303 (print) | NLM WE 103 | DDC
 612.7/6--dc23
LC record available at https://lccn.loc.gov/2017054181

ISBN: 978-1-4925-5843-9 (print)

This book is a revised edition of *Sport Mechanics for Coaches, Third Edition*, published in 2010 by Human Kinetics, Inc.

The web addresses cited in this text were current as of March 2018, unless otherwise noted.

Senior Acquisitions Editor: Joshua J. Stone
Senior Developmental Editor: Melissa Feld
Managing Editor: Kirsten E. Keller
Copyeditor: Bob Replinger
Indexer: Ferreira Indexing
Permissions Manager: Martha Gullo
Senior Graphic Designer: Nancy Rasmus
Cover Designer: Keri Evans
Cover Design Associate: Susan Rothermel Allen
Photograph (cover): Getty Images/Simonkr
Photographs (interior): © Human Kinetics, unless otherwise noted
Photo Asset Manager: Laura Fitch
Photo Production Manager: Jason Allen
Senior Art Manager: Kelly Hendren
Illustrations: © Human Kinetics, unless otherwise noted
Printer: Walsworth

Printed in the United States of America   10 9 8 7 6 5 4 3

The paper in this book was manufactured using responsible forestry methods.

Human Kinetics
1607 N. Market Street
Champaign, IL 61820
USA

United States and International
Website: **US.HumanKinetics.com**
Email: info@hkusa.com
Phone: 1-800-747-4457

Canada
Website: **Canada.HumanKinetics.com**
Email: info@hkcanada.com

E7186 (paperback) / E7292 (loose-leaf)

Tell us what you think!
Human Kinetics would love to hear what we can do to improve the customer experience. Use this QR code to take our brief survey.

To those on a quest for knowledge on sport mechanics and the practical application of science in sport.

CONTENTS

PART II   Putting Your Knowledge of Sport Mechanics to Work

PREFACE

The fourth edition of *Applied Sport Mechanics* is a comprehensive and innovative text on sport mechanics tailored for undergraduate students. This new edition is built on the fundamental laws that influence how humans move and perform in sport. Student learning of kinetics, kinematics, and sport technique are enhanced because these foundational principles are presented and discussed in a variety of applied scenarios.

The underlying theme is to enhance understanding of how humans perform—that is, how they respond to the stresses or loads from sport and exercise. The new knowledge generated by understanding this process can naturally be applied to nonsport scenarios, such as rehabilitation within the health industry.

This text doesn't concentrate on any one particular sport, such as football rowing, or basketball, or on any specific sport skill, such as punting, passing, serving, or spiking. Instead, *Applied Sport Mechanics* explains how and why a basic understanding of mechanical principles helps produce improved sport performance.

Organization

The format of this edition has been modified to include succinct chapters that can align with a typical undergraduate course format. The structural layout for the 13 chapters of the text, which reflects a typical teaching term, comprises two parts. Part I contains 10 chapters and describes the foundations of movement in sport, and each chapter contains a practical section at the end. The key elements of the chapters are sport mechanics anatomy (chapter 2), sport mechanics fundamentals (chapter 3), linear motion (chapter 4), linear kinetics (chapter 5), angular motion (chapter 6), angular kinetics (chapter 7), stability and instability (chapter 8), sport kinetics (chapter 9), and moving through fluids (chapter 10).

In part II, on the application of sport mechanics, the popular themes in a range of sports remain, with some enhancements. This section forms a handy quick reference guide for sport analysis.

Chapters 11, 12, and 13 provide more examples on how you can apply the information you learned in part I. In essence these chapters put your knowledge of applied sport mechanics to work. These chapters discuss why athletes must make their muscles work as a team and why synchronizing and coordinating muscle actions is important. This synchronization and coordination produces superior techniques that have resulted in world r ecords.

Features

The new edition also contains specific features within each of the chapters. These features have been developed to enhance student learning and understanding of sport mechanics.

- *Technology:* Keeping pace with today's lifestyle, the fourth edition has expanded the technology sections that were initiated in the previous text. The application of relevant technology can provide new insights into what is happening in sport.

- *A range of topics and applications:* To connect better with students, several modalities are used to explain and describe the human–sport interaction within each chapter. These scenarios can stimulate the understanding of the fundamental mechanical principles, which can then be applied to a range of situations.

- *At a Glance:* A new summary section in each chapter helps consolidate the key messages for the topic at hand. These reviews can assist student learning.

- *Practical examples:* The practical examples of how a player and coach currently integrate sport mechanics within their training and game analysis can open the door for the undergraduate student.

- *Application of knowledge:* The Application to Sport section has also been expanded to include examples within each chapter and relevant discipline area.

- *References:* To reflect industry practice and assist future professionals, peer-reviewed references are provided within each chapter. These references substantiate the applied nature of the material and provide a window for students to seek further knowledge on the various topics.

Student Resources

This edition comes with a web resource for students that contains review questions and practical activities. To access the web resource, follow the instructions described in the first page of the book. The web resource may be accessed at www. HumanKinetics.com/AppliedSportMechanics.

Instructor Resources

For this new edition a range of ancillary materials has been created. These materials include an instructor guide, presentation package and image bank, and a test package. The fundamentals of measuring mechanical aspect within selected sports can form the basis for practical activities that provide students with hands-on experience.

Human Kinetics and I believe that *Applied Sport Mechanics* will become the go-to resource for students and professors alike as they investigate and expand their knowledge of sport mechanics.

ACKNOWLEDGMENTS

I would like to acknowledge my colleagues, Dr. Mark Sayers and Dr. Luke Hogarth, for their valuable contributions to this text.

PART I

Foundations of Movement in Sports

1

Introduction to Applied Sport Mechanics

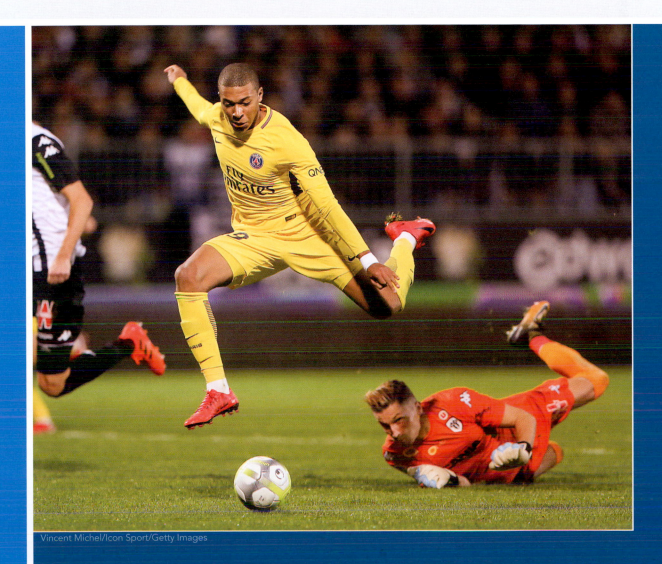

Vincent Michel/Icon Sport/Getty Images

When you finish reading this chapter, you should be able to explain

- what applied sport mechanics is;
- how the human body moves;
- how and why a basic understanding of mechanical principles helps produce an improved performance;
- how mechanical principles apply to sport;
- how to use sport mechanics to assess sport technique; and
- how *Applied Sport Mechanics* is organized so that you can gain the most knowledge from this topic.

For an undergraduate student, understanding how the human body moves is fundamental for a future career in sport, exercise, or the health care industry. In this textbook we combine the mechanical knowledge with applications in sport. By presenting in this perspective the aim is to provide a different pathway for gaining new knowledge in sport **mechanics**. *Applied Sport Mechanics* is written for undergraduate students, coaches, and teachers, but the fundamental knowledge can be applied by sport scientists, athletes, and sport fans. The book tells you how knowledge of sport mechanics helps produce better performances. Those of you who coach will find that this book helps you become a better coach. Those who assist coaches, namely sport scientists, will learn how to apply the mechanics of **human movement** to real-world conditions in sport. Both of these professions are careers that the sport mechanics graduate can aim for. Finally, athletes will discover information that helps improve their performances.

The underlying theme is to gain a better understanding of how the human body responds to the stresses or loads from sport and exercise. The new knowledge generated by understanding this process can naturally be applied to nonsport scenarios, such as rehabilitation within the health industry. A key mantra of recent times is "Exercise is medicine." Although this text has a specific focus on sport mechanics, the principles and **skill** sets developed are transferable to exercise rehabilitation.

Applied Sport Mechanics is also valuable to those who neither coach nor compete but who are enthusiastic sport fans. If you are a fan, you'll find that *Applied Sport Mechanics* helps you become a more critical and appreciative observer of the sports you love.

When you read *Applied Sport Mechanics*, you'll find that it doesn't concentrate on any particular sport or on any specific sport skill. Instead, *Applied Sport Mechanics* explains how and why a basic understanding of mechanical principles helps produce improved sport performance. From this scientific understanding you'll be able to look at an athlete's performance and say to yourself, "Some of the actions in the athlete's technique are good. But I can see actions that are inefficient and need correcting. What I know of mechanics tells me that they're wrong, and I know what kind of movements should replace them. When the corrections are made, the athlete will have a more efficient technique, show a safer movement pattern, and ultimately produce a better performance."

How Applied Sport Mechanics Is Organized

Applied Sport Mechanics is divided into two parts. Part I, containing chapters 1 through 10, explains the fundamentals of sport mechanics. Within these chapters a range of various applications of sport mechanics is embedded in the text, as are Application to Sport and At a Glance sidebars. Part II, which comprises chapters 11 through 13, delves further into the heart of *Applied Sport Mechanics*.

The chapters within part I cover the traditional concepts of the biomechanics of sport. Because biomechanics is all about the mechanical interaction of a human, the chapters naturally contain the following topics:

- Chapters 2 and 3: sport mechanics anatomy, followed by sport mechanics fundamentals
- Chapters 4 and 5: linear motion and linear kinetics in sport
- Chapters 6 and 7: angular motion and angular kinetics in sport

- Chapter 8: stability and instability
- Chapter 9: sport kinetics
- Chapter 10: moving through fluids

These chapters explain the interaction between an athlete, the equipment, and the ever-present external forces that assist or oppose the athlete during the performance of a sport skill. You'll learn what forces are at work when sprinters accelerate, when gymnasts spin in the air, and when pitchers hurl curveballs. You'll understand the mechanics of good technique and the advantages and disadvantages when athletes compete at high altitudes.

In part II, chapters 11, 12, and 13 provide more examples of how you can apply the information you learned in part I. These chapters discuss why athletes must make their muscles work as a team and why they need to synchronize and coordinate muscle actions. This synchronization and coordination produce superior techniques.

Chapters 11 and 12 are particularly useful to coaches, physical educators, and sport scientists because they explain how to observe and analyze an athlete's technique and how to correct errors. Each chapter gives you a series of steps to follow. You'll learn how to break down a skill into phases and what to look for as you analyze each phase. Then you'll read a series of important mechanical principles that you can refer to when you start correcting errors.

In chapter 13 you'll learn about techniques and mechanics in a wide range of sport skills, including sprinting, jumping, swimming, lifting, throwing, and kicking. First you'll read descriptions of the most prominent features in the performance of these skills. Then you'll learn the mechanical reasons why the technique of each skill is best performed in a certain way. The goal of this chapter is to show you how technique and mechanics are inseparable, no matter the sport.

Applied Sport Mechanics finishes with a glossary and a list of references that will help you expand your knowledge of sport mechanics. The glossary avoids dull, scientific explanations. Instead, it relates scientific principles to athletes and to the movement of sport implements such as bats, balls, and javelins.

Understanding how physical laws influence sport performances will help you, whether you are coaching, competing, or simply watching as a spectator. If you coach, remember that sport mechanics is just one tool that you'll use. You'll also need to increase your knowledge in such areas as sport psychology, physiology, nutrition, sport injuries, and the teaching of sport skills.

After you've read *Applied Sport Mechanics*, you should be able to watch an athlete perform and immediately know what external forces the athlete must contend with. You'll be able to analyze an athlete's movements and immediately recognize how they can be improved. You'll spot poor actions and replace them with efficient movements that produce quality technique based on sound mechanical principles.

APPLICATION TO SPORT

Surprising Facts on World Records

- Some records have been improved with modern-day technology, whereas others have remained for over 30 years.

- In pole vaulting, the world record was set in 2014 at 6.16 m (20 ft 2-1/2 in.), whereas the 2016 Olympic record is slightly less at 6.03 m (19 ft 9-1/4 in.).

- In high jumping, the women's world record was set in 1987 at 2.09 m (6 ft 10-1/4 in.), and the Olympic record from 2004 remains at 2.06 m (6 ft 9 in.).

- At the recent 2016 Rio de Janeiro Olympic Games, 27 new world records were set along with 91 Olympic records, from 306 medal events.

Pascal Rondeau/Allsport/Getty Images

What Is Sport Mechanics?

Sport scientists work in the field of **biomechanics**, a discipline that assesses the effects of forces like gravity and air resistance on humans and, conversely, the effects of forces that humans apply (Rice et al. 2010). The fundamental values of applied sport mechanics are derived from the basic laws of force and motion. More important, sport mechanics applies these laws to human movement, that is, sport or exercise.

The foundations of biomechanics that are developed in this textbook include **linear and angular kinematic analysis** (that is, defining position, displacement, velocity, acceleration), **linear and angular kinetics**, forces, laws of motion, center of mass, torque, and moments of inertia. Each of these standard concepts is discussed within the following chapters.

Obviously, gravity and air resistance, or even the forces that occur during collisions, make no distinction between nonsporting and sporting activities:

- A high jumper fights against gravity in attempting to clear the high-jump bar, just as a person climbing stairs or an airplane taking off has to overcome the forces of gravity.

- Likewise, air resistance opposes both an automobile and an Olympic sprint cyclist.

These examples tell us that the same mechanical principles used in our everyday world can be applied to sport. A typical example of biomechanics is shown in figure 1.1, in which the forces generated are overlaid on top of the human anatomy. Combining this mechanical principle with human anatomy allows you to understand sport mechanics.

Mechanical Principles

In sport, mechanical principles are nothing more than the basic rules of mechanics and physics that govern an athlete's actions. For example, if a coach and an athlete understand the characteristics of the earth's gravitational force, they know what must be done to counteract the effect of this force and, conversely, what actions must be performed to make use of it. Here are some typical examples:

- A springboard diver who is aware that gravity acts perpendicularly to the earth's surface has a better understanding of the trajectory that gives an optimal flight path for a dive.

- Similarly, wrestlers quickly learn that gravity is their friend when they have their opponents off balance. On the other hand, if they don't maintain their own stability, gravity can team up with their opponents!

- Ski jumpers understand that if they tuck their body into a more streamlined shape by flexing their legs and bend forward as they accelerate down the in-run, they reduce air resistance. This body position allows them to increase their maximum potential acceleration and go faster in preparation for the takeoff. Takeoff velocity is a key factor for projectile motion. We'll discuss that more in chapter 4, Linear Motion in Sport. Once in flight, ski jumpers counteract the force of gravity by making use of air resistance. They extend their legs and lean forward to deflect air downward. In response, the air pushes them upward. This varied use of gravity and

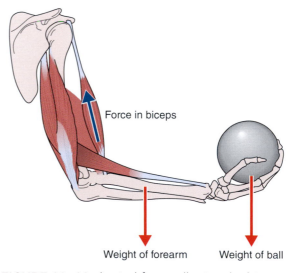

Force in biceps

Weight of forearm   Weight of ball

FIGURE 1.1 Mechanical forces allowing the biceps to hold up a ball.

Adapted from S.J. Hall, *Basic Biomechanics*, 4th ed. (Boston: McGraw-Hill, 2003), 158.

air resistance helps Olympic ski jumpers fly to distances in excess of 130 m (426 ft).

Many forces besides gravity and air resistance operate on earth. These forces act in different ways, and if you're in a contact sport you must also consider the forces produced by your opponents. As a student coach, the more you understand how all these forces interrelate, you'll be better able to analyze an athlete's technique and improve the athlete's performance. If you are an athlete and you have this knowledge, you'll understand why it's better to apply muscular force at one instant than at another and why certain movements in your technique are best performed in a particular manner. Even as a spectator and sport fan, you'll find that understanding basic mechanical principles can help you become more knowledgeable and appreciative of what it takes to produce an excellent performance.

In sport, the laws of mechanics don't apply to the athlete alone. Mechanical principles are used to improve the efficiency of sport equipment and playing surfaces. Modern track shoes, speed skates, skis, slick bodysuits for swimming and cycling, and safety equipment (such as pole-vault landing pads) are all designed with an understanding of the external forces that exist on earth and the forces that an athlete produces. This knowledge has been instrumental in raising the standard of performance in every sport.

AT A GLANCE

Sport Mechanics Measures

When looking at sport mechanics terminology, one of the key delineators is determining how to describe and measure the mechanics of the sport.

- If the description and measurement use numbers, such as reporting on distance in **meters** (feet), this account is a **quantitative** measure.

- On the other hand, if the report uses terms like "better" or "worse," or "smooth" or "clunky," these words are describing the quality of the movement and are therefore a **qualitative** evaluation.

- The rudimentary quantitative measures are linear and angular kinematics, and linear and angular kinetics.

Either of these describe the movement, or more important, the technique, which is described in the next section.

How Applied Sport Mechanics Can Help You

Most people involved in sport, like coaches or support staff, are reluctant to study sport mechanics. From experience they know that it has meant tackling dry, boring texts loaded with formulas, calculations, and scientific terminology. These texts are frequently written by academics who fail to explain the relationship of good technique to the principles of mechanics in a manner that is meaningful to coaches and sport enthusiasts and that will connect with the undergraduate student.

You'll be happy to find that *Applied Sport Mechanics* is a different type of book. This book contains few formulas or calculations, and it uses familiar measurements in both imperial and **metric system** units. As a student whose career aspiration is to be a coach or sport scientist, to perform as an athlete, or to watch as a fan, *Applied Sport Mechanics* is a book that you can learn from immediately.

To get the most from *Applied Sport Mechanics*, all you need is a desire to know how and why things work in the world of sport. In other words, if you have curiosity and a desire to improve, you'll get a lot of useful information from this text. The new knowledge gained from understanding *Applied Sport Mechanics* forms part of the recognized qualifications in the sport and exercise domain. To become an accredited sport and exercise scientist, you will need to demonstrate your competency in sport mechanics, as well as in other specific disciplines such as physiology, psychology, and nutrition. Here's how:

You will learn to observe, analyze, and correct errors in performance. This benefit is the most important one you'll get from reading *Applied Sport Mechanics*. This text will help you develop a basic understanding of mechanics, and by using this knowledge you will be able to distinguish between efficient and inefficient movements in an athlete's technique. The information in this text will help you pick out unproductive movements and follow up with precise instructions that help optimize performance.

1. You won't waste time with vague advice like "Throw harder" or "Try to be more dynamic." Obscure tips like these only confuse and frustrate the athletes you're trying to help.

2. One of the most effective ways to observe, analyze, and correct errors is through the use of technology, so we've added a section on technology within each chapter to provide you with some additional resources.

On the other hand, if you're the athlete and your coach is not present, a basic knowledge of sport mechanics will help you understand why you should eliminate certain movements in your technique and instead emphasize other actions.

You'll be better able to assess the effectiveness of innovations in sport equipment.

When Greg LeMond of the United States won the Tour de France by a few seconds over Laurent Fignon of France, he certainly illustrated the value of fitness and determination. Equally as important, LeMond and the technicians who assisted him knew the importance of reducing wind resistance to a minimum, particularly during the Tour's final time trial. They realized that if LeMond could maintain a low, dart-like body position that allowed the air to flow smoothly over and past his body, he would spend less energy pushing air aside and could spend more energy propelling himself at high speed.

This knowledge of mechanics paid off! The same principles apply in other sports. You need to know what is gained from design changes in such items as golf clubs, tennis rackets, skis, speed skates, mountain bikes, and swimsuits.

Applied Sport Mechanics cannot teach you all there is to know in the world of sport equipment design, because changes and modifications will continue to occur at an ever-increasing pace. But this text will certainly give you a foundation of knowledge on which you can build.

You'll be better prepared to assess training methods for potential safety problems.

Think of an athlete squatting with a barbell on his shoulders. Where should the athlete position the bar? Should it be placed high on the shoulders or lower down? And what about the angle of the athlete's back during the squatting action? What are the mechanical implications of a full squat compared with half and three-quarter squats? How fast should the athlete lower into the squat position?

1. If you know about levers and torque, you'll understand why it's dangerous to bend forward when you squat. Likewise, if you are familiar with the characteristics of momentum and understand how every action has

---

APPLICATION TO SPORT

Advances in Equipment Require Changes in Technique

In the 1998 Nagano Olympics, world records in speed skating were continually broken by athletes using clap skates. The blade on a clap skate is hinged at the front of the shoe but not at the heel. The hinge allows the blade to stay in contact with the ice longer so that the skater is able to thrust at the ice for a longer time. The characteristic clapping noise occurs at the end of each stroke when the blade "claps" back into contact with the heel of the shoe. The clap skate requires a new technique in which the skater must push more directly backward and from the toe rather than out to the side and from the heel. Athletes who were accustomed to the older skates (with the blade permanently joined to the boot at heel and toe) had to adapt to the new equipment and change their technique.

DAVE BUSTON/AFP/Getty Images

Adapted from "The Athletic Arms Race" by Mike May in *Scientific American* special issue: "Building the Elite Athlete," Nov. 27, 2000, page 74.

an equal and opposite reaction, you'll know that dropping quickly into a full squat puts tremendous stress on the lower back, knees, and hips. You may have been teaching good technique but don't fully understand why one way of performing the technique is potentially dangerous and another is not. *Applied Sport Mechanics* will give you the reasons.

2. In gymnastics you will frequently see spotting techniques that provide a high level of safety, and you'll also see other techniques that endanger both the gymnast and the spotter.

To explain these concepts further, we've added a new section, "How to Measure," where appropriate within the relevant chapter. This new section describes several ways to measure what is happening in sport that can then be used to assess training methods. By reading *Applied Sport Mechanics* you'll discover, for example, why efficient spotting requires an understanding of balance, levers, torque, and the momentum generated by the gymnast performing the skill.

This information will help you teach safe spotting techniques in gymnastics and good technique in weight training. Of course, *Applied Sport Mechanics* is not limited to these two sports. You can apply the mechanical principles that you read about to every sport.

You'll be better able to assess the value of innovations in the ways sport skills are performed.   In sport, our capacity for reasoning and creativity has been responsible for the advances in talent selection, technique, training, and equipment design. We all possess a tremendous capacity for creativity. Coaches must use this creativity to search for better ways to improve their athletes' performance. All athletes differ in physique, temperament, and physical ability; what works for one athlete will not necessarily work for another. Similarly, a young athlete will differ dramatically from a mature athlete.

1. To help athletes achieve top-flight performances, coaches should learn why sport techniques are performed as they are and be prepared to modify certain aspects of these techniques to fit the athletes' age, maturity, and experience. There are many examples of the willingness of coaches and athletes to try out new ideas.

2. In team games, coaches modify attack and defense formations relative to the team they will face in an upcoming contest.

Among athletes, think of how the creativity and experimentation of Dick Fosbury revolutionized the high jump. Similarly, the glide and rotary techniques have increased the distances thrown in the shot put. In gymnastics, consider the number of skills named after their creative inventors (such as the "Thomas Flair," named after former U.S. gymnast Kurt Thomas).

So be curious and learn the how and why of technique. At the same time, be creative and willing to experiment. Coaches should to encourage their athletes to use their own creative capacities as well. They should always look for ways to improve their understanding of the sport they are coaching. They can be a coach, an analyst, and an innovator all at the same time!

You will know what to expect from different body types and different levels of maturity. If you understand the mechanical principles governing the techniques of your sport, you'll understand why young athletes (who are growing quickly) have a tougher time maneuvering, changing direction, and coordinating their movements than more mature athletes do.

1. You'll realize that young athletes cannot follow the same training regimens used by more mature athletes.

2. You'll also understand why tall athletes with long arms and legs have an edge in some sports but are at a disadvantage in others.

Similarly, you will realize why smaller athletes tend to have a good strength-to-weight ratio and can cut, turn, and shift more quickly than athletes who are taller and heavier.

Understanding and Quantifying Technique

When we compare the performances of two athletes, we often say that one of the two athletes has better form or, more precisely, that one athlete has better technique than the other. By **technique**, we mean the pattern and sequence of movements that athletes use to perform a sport skill, such as a forearm pass in volleyball, a hip throw in judo, or a somersaulting dive from the tower. Elite athletes create the image of having all the time in the world to complete a complex task. In biomechanical terms they have effectively refined their kinetic link—the timing of movement between successive limbs, like the thigh, shank, and foot during kicking (Sanders 2007).

Sport skills vary in number and type from one sport to the next. In some sports (e.g., discus and javelin), only one skill is performed. A discus thrower spins and throws the discus—nothing more. As such, this activity is defined as a closed skill; that is, it is closed to the thrower because the thrower is the person who determines when to start to spin and when to release the discus.

In tennis, in contrast, players perform forehands, backhands, volleys, and serves, and they perform all these actions in response to what the opponent has done. These actions are defined as open skills. Each skill, whether it's a tennis serve or a discus throw, has a specific objective determined by the rules of the sport. In a serve, a tennis player wants to hit the ball over the net and into the service area in such a way that the opponent cannot return it. A discus thrower aims to throw the discus as far as possible, making sure that it lands in the designated area. Both athletes try to use good technique so that they achieve the objectives of each skill with the highest degree of efficiency and success. Knowledge of sport mechanics is required to understand this process.

Applying Good Technique

An athlete can perform a skill with good or poor technique. Poor technique is ineffective and fails to produce the best results. In the worst-case scenario, poor technique can cause injury to the athlete and possibly to the people surrounding the activity.

You can see plenty of poor technique at any public driving range and, along with poor technique, inferior results! Hooks and slices are mixed in with wild swings that miss the ball entirely. Even if you know little about golf, you'll be amazed by the variation you see in the performances of a single stroke. Now compare these recreational golfers with elite professional players. Although elite players differ in height, strength, and weight, the basic technique they use in their strokes is much the same (Nesbit and McGinnis 2009).

From backswing to follow-through, a smooth application of force that appears graceful and fluid is what you're likely to see. This efficiency of motion tells you that elite golfers use good

---

Never Discount Individual Creativity and Inventiveness

Just as Dick Fosbury revolutionized high jumping, Graeme O'Bree from Scotland set the cycling world talking with his self-designed bike. O'Bree removed the bike's crossbar, shortened the bike's length, and made its profile as narrow as possible. Without a crossbar, O'Bree's legs brushed against each other as he cycled. His chest lay flat on top of the handlebars. O'Bree's body position reduced air resistance to a minimum. Although lying with the chest on the handlebars was subsequently declared illegal by the Cycling Federation, O'Bree was not to be beaten by the rules. He modified his bike even further so that he could cycle without having his chest contact the handlebars, but his upper body was still stretched forward horizontally like an arrow in what was called the Superman position. O'Bree became a world champion in the velodrome and knew the importance of reducing air resistance. Although the Superman position is illegal in the mass-start stages of the Tour de France, athletes attempt to cycle with their upper bodies in a horizontal position during the Tour's time trials. Reducing air resistance in these high-speed time trials is essential.

Mike Powell /Allsport/Getty Images

technique, with a refined kinetic link. They practice for hours to hone this technique so that their actions become highly effective and get the job done.

Apart from minor differences, all top-class athletes, whatever their sport, use superior technique based on the best use of the mechanical principles that control human movement. But remember that the refined, polished movements you see in the technique of an elite athlete seldom occur by chance. They usually result from hours of practice and, more important, smart practice—the right type of practice.

Nowadays, an athlete is extremely unlikely to reach world-class status without the assistance of coaches and sport scientists who know why it's better to perform movements one way rather than another. Today's top athletes get help from knowledgeable coaches who critically observe their performances and tell them which movements are efficient and which are not. As a student of sport you have a great opportunity to add your new knowledge of applied sport mechanics to the coach's toolbox. The coach's knowledge and **sport science** assessment, coupled with the athlete's talent and discipline, help produce safe, first-rate performances.

Teaching Good Technique

What must you know to teach good technique? As an example, let's look at what is necessary when you teach a novice to drive a golf ball. When you introduce this skill, being able to demonstrate good technique is an asset, although this ability is certainly not essential. What is more important is being able to analyze and correct faults in the novice's performance and use good teaching progressions to lead the novice to a more refined performance. To do this, you need a basic understanding of the mechanics of the golf drive, which means that you must know why certain actions in a golf drive are best performed in one way and not another.

The same idea is important when you coach or instruct classes in volleyball. A volleyball coach needs to know the mechanical reasons why certain movements get a player up in the air for a spike and why other movements do not (Markou and Vagenas 2006). In baseball, a pitching coach aims to teach the most efficient actions for the windup, delivery, and follow-through. Similarly, a batting coach works to make the batter a more effective hitter.

The golf, baseball, and volleyball coaches use their knowledge of mechanical principles when they eliminate inefficient movements and replace them with actions that are more efficient. One of the best ways to transfer this knowledge is to use a variety of processes to communicate with the athlete. An assortment of new technologies can be used to provide feedback to the athlete.

Failure of Traditional Training Methods and Use of New Technologies

Many coaches and athletes still follow traditional methods during training. They reason, "This is how it was done in the past and it worked well, so this is how we should do it now." Some coaches have no idea why some movements are good and others bad, why some are safe and effective whereas others cause injury and erratic actions. Other coaches are happy using a trial-and-error method. Without knowledge of sport mechanics, coaches will guess at the best way to correct or improve a technique. Sometimes they get good results, but more often they don't. This potentially misguided experimentation with an athlete could result in injury. Many coaches teach their athletes a technique based on a world champion's technique without taking into account differences in physique, training, and maturity.

Similarly, young athletes often copy every action of a world-class performer, including idiosyncrasies that are mechanically ineffectual. For example, Al Oerter, four-time Olympic discus champion from 1956 to 1968, frequently inverted the discus as he swung his arm back during his windup. This action was simply a personal trait that added nothing to the mechanical efficiency of Oerter's throwing technique, yet many young athletes copied it, believing that it would add distance to their throws. A more comical example of copying idiosyncrasies was seen in the number of kids who attempted slam dunks with their mouths open and tongues hanging out. Why? Because this was a common characteristic of Michael Jordan!

Being able to distinguish between safe, mechanically correct movements and those that serve no purpose is essential for skill development. Coaches and athletes who blindly mimic the methods and techniques of others progress only so far. Perhaps this limited focus on athletic technique is a reflection of the peer-reviewed publications in the sport science area. In a review of sport science research, the discipline of physiology—the study of how exercise or sport alters the structure and physiological function of the human body—was the most researched scientific discipline (Federolf et al. 2014). Because sport mechanics is the study of how the human body moves, this discipline is the best to describe technique. *Applied Sport Mechanics* will help you eliminate this haphazard approach. By developing a basic understanding of mechanics, you'll be able to analyze performances and teach movement patterns that produce efficient technique. This approach will lead to better performances.

AT A GLANCE

Sport Technique

- Technique is best described as the pattern and sequence of movements that athletes use to perform a sport skill.

- Although elite players differ in height, strength, and weight, the basic technique they use in their movements is much the same.

- Being able to distinguish between safe, mechanically correct movements and those that serve no purpose is essential for skill development.

Applying New Methods and Technology

Technology is part of everyday life, and some of this technology has a place within sport. For example, people use a global positioning system (GPS) in their cars or when hiking to monitor how far they have traveled or how long it will take to arrive at the destination (Townshend et al. 2008). Similar technology can be used in sport to determine how far an athlete has run during a game or to quantify how straight someone paddled a kayak. This modern-day knowledge can provide new feedback to athletes and thus tell them something they didn't already know. Or at times, this technology can tell them something that the coach has repeatedly said but that somehow didn't get through.

How many times has a golf coach told a golfer, "You are lifting your head just as the club is making contact with the ball"? The golfer doesn't feel her head moving and for a number of reasons doesn't or can't make a correction to her technique.

- If the coach uses a video camera to record the golfer's swing and then plays the video back, the golfer can see the fault in the mechanics.

- More advanced technology can measure the amount of movement, such as how far the head is lifted, and force transducers and EMG (electromyography) can measure the timing of muscle sequences.

This type of monitoring with technology can provide a new approach to coaching; therefore, in understanding sport mechanics, we need to understand and incorporate technology to correct and optimize movement and sport skill.

SUMMARY

- *Applied Sport Mechanics* provides a scientific understanding to explain how and why mechanical principles help produce improved performance.

- The field of biomechanics is a discipline that assesses the effects of forces on humans and, conversely, the effects of forces that humans apply. Biomechanics is the aspect of science concerned with the basic laws of force and motion.

- The foundations of biomechanics include linear and angular kinematic analysis (that is, defining position, displacement, velocity, acceleration), linear and angular kinetics, forces, laws of motion, center of mass, torque, and moments of inertia.

- In sport, mechanical principles are nothing more than the basic rules of mechanics and physics that govern an athlete's actions.

- In sport, the laws of mechanics don't apply to the athlete alone. Mechanical principles are used to improve the efficiency of sport equipment and playing surfaces.

- If the description and measurement use numbers, this account is a quantitative measure.

- On the other hand, if the report uses descriptive terms, these words are describing the quality of the movement and are therefore a qualitative evaluation.

- By technique, we mean the pattern and sequence of movements that athletes use to perform a sport skill.
- As a student of sport you have a great opportunity to bring your new knowledge of applied sport mechanics to the coach's toolbox.

KEY TERMS

biomechanics

human movement

linear and angular kinematic analysis

linear and angular kinetics

mechanics

meter

metric system

qualitative

quantitative

skill

sport science

technique

REFERENCES

Federolf, P., R. Reid, M. Gilgien, P. Haugen, and G. Smith. 2014. "The Application of Principal Component Analysis to Quantify Technique in Sports." *Scandinavian Journal of Medicine and Science in Sports* 24 (3): 491-99.

Markou, S., and G. Vagenas. 2006. "Multivariate Isokinetic Asymmetry of the Knee and Shoulder in Elite Volleyball Players." *European Journal of Sports Science* 6 (1): 71-80.

Nesbit, S.M., and R. McGinnis. 2009. "Kinematic Analyses of the Golf Swing Hub Path and Its Role in Golfer/Club Kinetic Transfers." *Journal of Sports Science and Medicine* 8:235-46.

Rice, I., F.J. Hettinga, J. Laferrier, M.L. Sporner, C.M. Heiner, B. Burkett, and R.A. Cooper. 2010. "Biomechanics." In *The Paralympic Athlete: Handbook of Sports Medicine and Science*, 31-50. Oxford, UK: Wiley-Blackwell.

Sanders, R.H. 2007. "Kinematics, Coordination, Variability, and Biological Noise in the Prone Flutter Kick at Different Levels of a 'Learn-to-Swim' Programme." *Journal of Sports Sciences* 25 (2): 213-27.

Townshend, A.D., C.J. Worringham, and I. Stewart. 2008. "Assessment of Speed and Position During Human Locomotion Using Nondifferential GPS." *Medicine and Science in Sports and Exercise* 40 (1): 124-32.

www Visit the web resource for review questions and practical activities for the chapter.

2

Sport Mechanics Anatomy

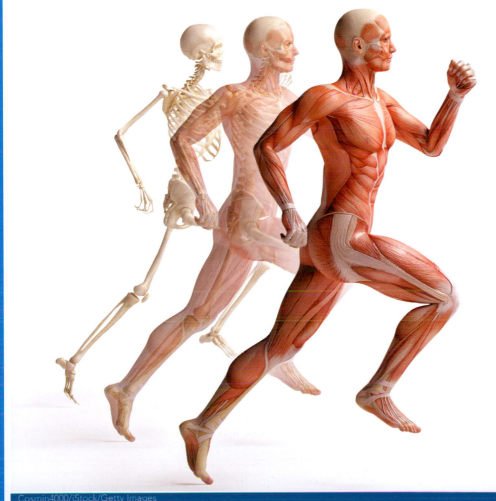

Cosmin4000/iStock/Getty Images

────────────────

When you finish reading this chapter, you should be able to explain

- anatomical movement and the anatomical reference position;
- the standard reference system and terminology;
- directional terms for mechanics and the mechanics of anatomy;
- how athletes make use of first-, second-, and third-class levers;
- what a lever system means in sport;
- what is happening when analyzing human movement;
- discipline-specific terminology;
- the relationship between the body and any external object (like a sporting implement); and
- how levers work in the human body.

────────────────

Our bodies have a framework of more than 200 bones. These bones form the skeleton, which supports and protects our inner organs and provides attachments for our muscles, tendons, and ligaments. Our bones articulate with other bones at the joints. We have a variety of joints; depending on their design, they determine the type of movement that occurs. As far as sport mechanics is concerned, the most important joints are those in our hips, knees, ankles, elbows, wrists, shoulders, and spinal column—all of which allow considerable movement (Tortora et al. 2016). These joints form axes around which major parts of our bodies (called body segments) rotate, such as the forearm and upper arm when we swing a golf club.

As the title suggests, this chapter covers only the anatomical aspects as they relate to sport mechanics. The discussion starts with the anatomical reference system so that you can describe anatomical movement consistently. From this reference point the direction of anatomical movement can be defined. To understand how this movement has occurred, the connection with the internal musculoskeletal **lever** system is described. The next topic is the application of these connections through the human lever systems. These levers enable a torque to occur, which results in human movement. Because the focus of this text is applied sport mechanics, specific details on the names and orientation of the muscles and bones within the human body are not covered because they typically come under a human anatomy course.

Like any subject or topic, this discipline has some discipline-specific terminology. That vocabulary is what this chapter on sport mechanics anatomy and the following chapter on sport mechanics fundamentals are all about.

Standard Anatomical Reference Terminology

As a student of sport mechanics, you need to understand the common, universal system used to describe what is happening when analyzing human movement. That way you will be able to understand and connect with colleagues and other professionals as you apply your sport mechanics knowledge. The fundamental approach is to observe and describe human movement from

- the side view,
- the front view, and
- the top view.

These three planes of view allow a range of observations or measurements to be made. Ensuring that the observations or measurements are made in these specific planes of view controls for any measurement error, that is, inaccuracies due to the viewing angle. We'll look more at the measurement error in chapter 3.

Anatomical Reference System

To keep a consistent measurement process, we begin with an anatomical reference position. This reference system is the starting point for human movement analysis. It is not a typical or natural body position, but it does provide the best starting point for describing human movement. The anatomical reference system is defined as getting into the following body setup, as in figure 2.1:

- Standing erect
- All body parts facing forward, including the palms of the hands

1. Looking from the side view is known as observing movement in the sagittal plane.

2. Looking from the front view is known as observing movement in the frontal plane.

3. Looking from the top view is known as observing movement in the transverse plane.

Directional Terms

From this anatomical body position, the relationship between the body and any external object like a sporting implement can be defined. More important, the direction in which the human body moves can be described. The directional terms for describing human movement are shown in table 2.1.

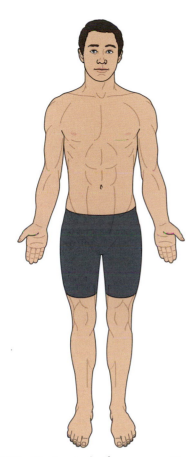

FIGURE 2.1   Anatomical reference position.

From this standard reference position we can observe human movement from the three views just described. Then from each of these different perspectives we create three specific cardinal reference planes. The reference planes are like a two-dimensional window and allow you to break down the movement into digestible chunks. The relationship between the view and the plane is as follows:

AT A GLANCE

Anatomical Reference Systems

- The three standard planes of view required before observing or measuring human movement are the front view, side view, and top view.

- The initial body setup of any anatomical reference system is defined as standing erect with all body parts facing forward, including the palms of the hands.

- The three specific cardinal reference planes are the sagittal plane, the frontal plane, and the transverse plane.

Connection Between Human Anatomy and Sport Mechanics

From this anatomical reference system we can look at how humans move, that is, the connection between human anatomy and sport mechanics. For

---

TABLE 2.1   Directional Terms for Sport Mechanics

Term	Descriptor
Anterior	Toward the front of the body (the front)
Posterior	Toward the back of the body (the back)
Medial	Toward the midline of the body (the middle)
Lateral	Away from the midline of the body (the outside)
Superior	Toward the head (the top)
Inferior	Toward the feet (the ground)
Proximal	Closer to the relevant anatomical joint (closer to the human joint)
Distal	Farther from the relevant human joint (away from the human joint)

humans, bones rotate at the joints, so an athlete's muscles, bones, and joints work together like a mechanical lever system. We will investigate this topic in more detail in the following section. In essence, human muscles pull on the bones (which act as the force to move the **resistance**), and the bones rotate at the joints. When causing this movement there are two types of muscle actions: **agonist** (a muscle contraction to help the movement occur) and **antagonist** (a muscle contraction that opposes the direction of movement). The joints act as the axis (**fulcrum**, or pivot point), like a hinge on a door. All human movement is a consequence of force and resistance battling each other.

- Force is applied at one location on the lever, and a resistance applies its own force at another. The action of the applied force attempts to make the lever rotate in one direction.

- The resistance then tries to make the lever rotate in the opposing direction.

In an athlete's body, force is primarily produced by muscular contraction. The **weight** of the athlete's limbs, plus the weight of whatever the athlete is trying to move, produces resistance. Figure 2.2 shows the position of force and resistance in a dumbbell curl, which is an exercise that most athletes use.

In addition, in a number of situations in sport we can manipulate the lever length, such as by shortening (or choking) the grip on the baseball

bat or using a pitching club instead of the longer driver in golf. These examples show that changing the grip (the human anatomy) can then change the swing path of the bat or club (the swing mechanics). The result is the transfer of forces from the human to the bat or club to the ball (Nesbit and McGinnis 2009).

Another example is the performance of a biceps curl exercise, as shown in figure 2.2. When the athlete contracts the biceps to generate the curl movement, because the biceps muscle is attached to the forearm, the muscle contraction is right near the elbow joint. At that muscle attachment point the lever moves in an arc of, say, 1 unit in length. But if the dumbbell is held farther from the elbow joint, it travels around an arc that is, say, 10 times larger. This increase in range of movement, shown schematically in figure 2.3, occurs because the **resistance arm** is 10 times longer than the force arm. Furthermore, if it takes 1 s for the athlete's biceps to complete its arc of 1 unit, the dumbbell moves through an arc 10 times larger in 1 s as well. Thus, the longer the athlete's arm is, the faster and farther the dumbbell moves.

These characteristics illustrate the compromise we all experience because of the way our bodies are designed. We are all at a disadvantage because our muscles must exert tremendous force to move a much lighter resistance. Obviously, if we don't possess the necessary strength, we cannot move even a light resistance. And to make matters worse, the longer our limbs are, the greater the force required from our muscles is.

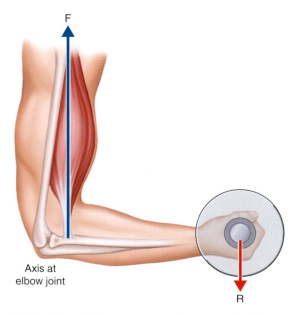

Axis at elbow joint

FIGURE 2.2   A lever system in the human body.

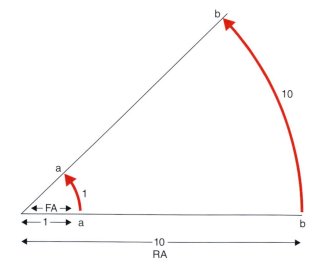

FIGURE 2.3   A mechanical representation showing the relationship between arc length and distance traveled. Point b moves 10 times farther and 10 times faster than point a on the force arm.

Torque

As the human bones rotate at the joints, they always produce a turning effect, which we call **torque**. In auto mechanics, a torque wrench is designed to apply a precise turning effect to a bolt. In weight training, a dumbbell curl requires the biceps muscle to pull on the forearm and produce a turning effect in an upward direction. How much torque occurs depends on the amount of force produced by the biceps, multiplied by the length of the force arm. For an athlete performing a dumbbell curl, the force arm is the perpendicular distance from the biceps' attachment on the forearm to the axis (i.e., the elbow joint), as shown in figure 2.4. The dumbbell and the weight of the athlete's forearm generate their own torque as gravity pulls them down.

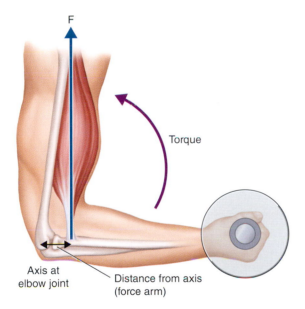

FIGURE 2.4 Torque is force multiplied by the perpendicular distance from the axis to where the force is applied.

To understand how we can increase the turning effect of torque, let's have an athlete become a mechanic and use a wrench to loosen a bolt. When the athlete pulls on the wrench, he applies torque to the bolt. Whether the athlete is successful in overcoming the rotary resistance of the bolt depends on how much force he exerts, how far from the bolt he applies force (i.e., the length of the force arm), and angle at which he pulls on the wrench. A 90° angle of pull is most efficient. In figure 2.5a, the athlete applies 10 units of force at 5 units of distance from the bolt (which acts as the axis). The torque produced is 50 units. In figure 2.5b, the athlete applies the same amount of force 10 units of distance from the bolt.

- In this case the turning effect, or torque, applied to the wrench has been doubled to 100 units because the force arm has been made twice as large.

- If the athlete wants to produce even more torque, the options available are to apply more force at 90° to the wrench, to increase the size of the force arm, or to do both in combination.

From this example you'll notice that a large force arm is an important requirement if an athlete wants to produce a large amount of torque. Now let's look at an example of a baseball bat, as shown in figure 2.6. The principles followed by the athlete in applying torque to a bolt apply to all lever situations that occur in sport (Kageyama et al. 2014). In essence, what happens is a battle between the turning effect of two opposing torques—torque produced by an athlete's muscular forces and torque produced by any type of resistance, such as the weight of the athlete's limbs plus whatever the athlete is holding (e.g., a discus or a barbell). In a sport such as wrestling or judo, an opponent can also generate resistance and torque.

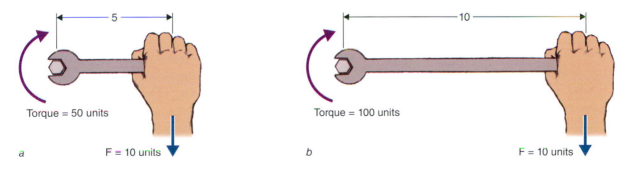

FIGURE 2.5 Torque is doubled when force is applied at twice the distance from the axis. (a) Force is applied 5 units from the axis, producing a torque of 50 units. (b) Force is applied 10 units from the axis, producing a torque of 100 units.

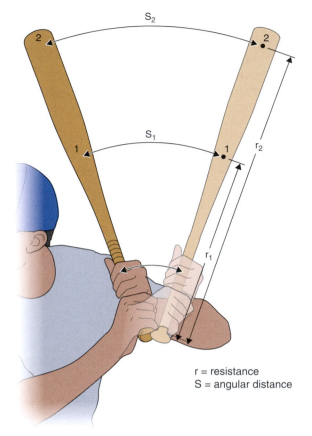

r = resistance
S = angular distance

FIGURE 2.6   An example of the relationship in a baseball bat, torque, and force. If the bat is swung at a constant velocity and the ball makes contact with the bat at point 1, the force on the ball would be, say, 50 units. But if the ball makes contact with the bat at point 2, which is twice the distance of r_1, the force on the ball is double, or 100 units.

Adapted from S.J. Hall, *Basic Biomechanics*, 4th ed. (Boston: McGraw-Hill, 2003), 370.

AT A GLANCE

Human Anatomy-Mechanics Connection

- For humans, bones rotate at the joints, so an athlete's muscles, bones, and joints work together and work like a mechanical lever system

- All human movement is a consequence of force and resistance battling each other.

- In an athlete's body, force is primarily produced by muscular contraction. The weight of the athlete's limbs plus the weight of whatever the athlete is trying to move creates resistance.

- Because the human bones rotate at the joints they always produce a turning effect, called torque.

Lever Systems

From this understanding of the term *torque*, we can then explore how various torque scenarios can exist within the human body. This mechanism is best described as a human lever system. Using the familiar bicycle pedal as an example, the mechanical principle for rotation requires an **axis of rotation** and a lever. For a lever system to work, a force is applied to the lever; if this force is greater than the resistance to rotation at the axis, angular motion will occur. If we go back to the bicycle pedal, force is applied by pushing down on the pedal; if this force is greater than the resistance, the crank will turn and the bicycle will move. If you are trying to pedal up a steep hill and cannot push down with enough force on the pedal, the crank will not move and the bicycle will not go up the hill.

To understand how levers work in the human body and to see what part they play in sport, let's first look at the components of a lever system, that is, the axis (or fulcrum), the force arm (or FA, which is typically the human muscle), and the resistance arm (or RA, which typically is what is to be moved). Some people may start to get nervous because they think they can't do mathematics, but levers are simple and require only the multiplication of two measures—how much force is being applied and at what distance the force is applied (this can be the effort required to move an object or the amount of resistance to move the **mass** or object). To show this, we use an example much like a seesaw on a playground (see figure 2.7).

1. To calculate how much force (F) is required for a lever system, we need to use the perpendicular distance (that way we know that all the force is being applied to move the lever) from the axis to where the force is applied; this is called the **force arm**, or just the lever arm, shown as the symbol FA in figure 2.7.

2. We need the perpendicular distance to calculate the moment, or torque (this will be explained in more detail in the next section). Likewise, the perpendicular distance from the axis to where the resistance (R) applies its own force is called the resistance arm, shown as RA in figure 2.7.

3. The axis (which is just the fulcrum, or pivot point) on which the lever rotates is generally in the center, and the lever is the rectangular box, or a seesaw's long seat. If two children of equal mass sit on each end of a seesaw, it will be balanced, meaning that the force and

resistance are equal. If one child is heavier than the other, naturally the lighter child will be left up in the air. The lighter and the heavier child act as the force and resistance arm, respectively.

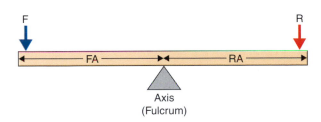

FIGURE 2.7   Components of a lever system, the axis (or fulcrum), the force arm (or FA, which is typically the human muscle), and the resistance arm (or RA, which typically is what is to be moved).

Types of Levers

Mechanically, levers are divided into three groups called first, second, and third class. This classification is based on how the force, resistance, and axis are positioned on the lever relative to each other. In an athlete's body, third-class levers are most common. In the performance of sport skills, however, you'll find that athletes frequently use all three classes of levers.

An easy way to remember the relationship of force, resistance, and axis for each of the three lever systems is to use the acronym *ARF*; *A* is associated with a first-class lever, *R* is associated with a second-class lever, and *F* is associated with a third-class lever.

1. On a first-class lever, *A* (the axis) is positioned between the resistance and the force.
2. On a second-class lever, *R* (the resistance) is positioned between the axis and the force.
3. On a third class-lever, *F* (the force) is positioned between the axis and the resistance.

First-Class Levers

In a **first-class lever**, the axis is positioned somewhere between the force and the resistance, as shown in figure 2.7. The force arm is the perpendicular distance from the axis to where force is applied to the lever, and the resistance arm is the perpendicular distance from the axis to where the resistance applies its force to the lever.

- Force and resistance arms can be equal in length, as they are in figure 2.7, or they can be unequal.

- If the force arm is longer than the resistance arm, then the lever favors force output, meaning that this lever arrangement multiplies the force exerted by the athlete. As a result, on the opposing side of the axis (fulcrum or pivot point), a greater force is applied to the resistance.
- If the force arm is shorter than the resistance arm, then the lever favors speed and range of movement at the expense of force.
- What is lost in force produces a gain in velocity and distance, and what is gained in force output occurs only with a loss in velocity and distance. So there's always a compromise.

To show how this compromise occurs in sport, we look at a weight-training machine, which frequently uses first-class levers in its design, as an example. Figure 2.8 shows a simplified illustration of a leg press machine. The athlete applies force with her legs on one side of the axis, and the weight stack as the resistance applies its own force on the other. On some leg press machines, the athlete will find two sets of foot pedals to push against. One set is purposely positioned lower on the machine than the other.

An athlete who struggles to lift the weight stack using the upper pedals will find it much easier using the lower pedals. Why?

- On the lower set of pedals, the force arm (which is the perpendicular distance from the axis of the machine to the line of force applied by the athlete's legs) is much longer than it is on the upper set.
- As a result, the same amount of force applied by the athlete's legs to the lower set of pedals produces greater torque.
- The turning effect of torque produced by the athlete in a clockwise direction must overcome the torque generated by the weight stack acting in a counterclockwise direction.

But if something is gained by the athlete in using the lower set of pedals, then something has to be lost or given up (McKean et al. 2010). The compromise in this situation is that the athlete must push the lower set of foot pedals through a bigger arc of movement than when using the upper set.

Second-Class Levers

A **second-class lever** is characterized by the force and resistance being on the same side of the axis and the force arm always being longer than the

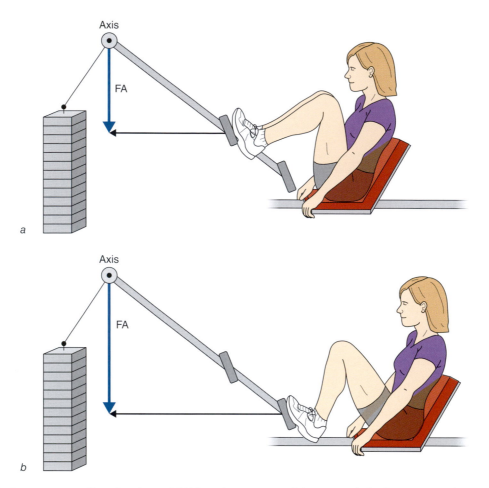

FIGURE 2.8 Leg press as a first-class lever. (*a*) When the upper pedals are used, the force arm is shortened and more force is required to raise the weight stack than in (*b*), when the lower pedals are used.

resistance arm (see figure 2.9). The applied force (which could be the force applied by you) acts in one direction (e.g., counterclockwise), and the resistance tries to move in the other direction (i.e., clockwise). If you apply sufficient force, then force and resistance move in the same direction. Second-class levers favor the output of force at the expense of velocity and range of movement. The larger the force arm is in relation to the resistance arm, the greater the force output is. An athlete who uses a second-class lever applies less force

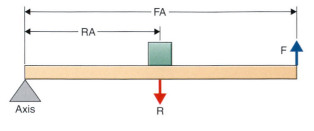

FIGURE 2.9 Second-class lever.

over a large distance at a greater speed to shift a heavier resistance a small distance at a slower velocity. The following example demonstrates these characteristics.

A second-class lever arrangement is shown in figure 2.10, in which an athlete performs a bench press on a multipurpose weight machine. Notice the following:

- The athlete pushes upward on handles attached to a single bar; the bar rotates at an axis to the left of the machine.
- The weight stack, supported by rollers that run on the upper surface of the same bar, is positioned closer to the axis than the handles on which the athlete is pushing.
- The force exerted by the athlete moves upward in an arc, and the rollers allow the weight stack to be lifted vertically.

When the athlete's arms are at full extension, the perpendicular distance of the force arm is

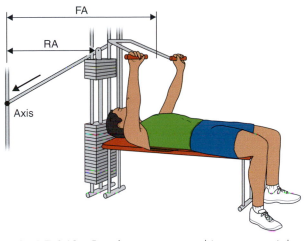

FIGURE 2.10 Bench press on a multipurpose weight-training machine, in which the human lifts the weight on the bench press, which operates as a second-class lever system.

reduced in length relative to the perpendicular distance of the resistance arm. In other words, as the athlete straightens his arms and becomes anatomically more efficient in applying force with the muscles of his arms, chest, and shoulders, the weight machine counters this anatomical change by reducing the length of the force arm relative to the resistance arm.

Third-Class Levers

In a **third-class lever**, the axis is at one end of the lever and the applied force is always closer to the axis than the resistance (see figure 2.11). As in the second-class lever, if the applied force is great enough to move the resistance, then applied force and resistance move in the same direction. This characteristic distinguishes second- and third-class levers from first-class levers, in which applied force and resistance move in opposing directions. Third-class levers always move the resistance through a larger range and at a greater velocity than moved by the force. On the other hand, the force applied is greater than that applied by the resistance.

A biceps curl is a great example of a third-class lever because it shows the typical relationship that exists between most of the muscles, bones, and

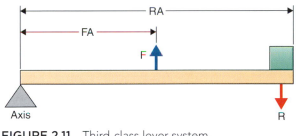

FIGURE 2.11 Third-class lever system.

APPLICATION TO SPORT

Olympic Pole Vault Differs From Dutch Canal Vaulting

One of the rules in pole vault says that a vaulter cannot shift the upper handhold. This rule was established because in the old days, vaulters were monkey-climbing the pole! If you visit Holland, you'll see a competition in which climbing the pole still occurs. By climbing up the pole the athlete is changing the length of the lever. The Dutch, for fun, pole-vault for distance across water-filled canals. Climbing the pole gives them more distance and a dry landing on the other side of the canal. If they climb the pole too soon or too high, they can stall and drop in the water, or even fall back

VALERIE KUYPERS/AFP/Getty Images

toward the takeoff. Spectators love the drama and the inevitable dunkings! When vaulters climb the pole, their rotation slows. A musician's metronome does the same thing. Raise the weight, and the metronome clicks at a slower tempo. Keep raising the weight, and the metronome stops altogether. The same principle applies to Dutch canal vaulting.

joints in an athlete's body. It also demonstrates that the muscles must exert considerable force to move even a light resistance.

In figure 2.12, an athlete is holding a dumbbell so that it neither rises nor falls. The counterclockwise turning effect (torque) produced by the athlete's biceps equals the torque produced by gravity acting in a clockwise direction on the mass of the athlete's forearm and the dumbbell. In this example we have set the ratio of resistance arm to force arm at 10:1.

- If the resistance is 10 units and the distance of the resistance to the elbow joint is 10 units, then the clockwise turning effect of the resistance is 100 units (10 × 10 = 100).

- If the force arm is only 1 unit of distance, the force necessary to hold the forearm and dumbbell horizontal is 100 units, or 10 times that of the resistance.

We must mention one additional characteristic here. With the arm held horizontal, the biceps normally pulls on the forearm at an angle slightly less than 90°. This arrangement helps rotate the forearm to the upper arm and also pulls the forearm toward the elbow joint, which helps to stabilize this joint (physiologically, this helps hold the bones at the elbow joint together).

In figure 2.13, we have purposely exaggerated the angle of pull of the biceps so that the division of the force produced by the biceps is more apparent. In this figure, the contraction of the biceps pulling the forearm toward the upper arm is approximately

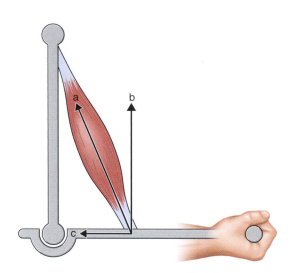

FIGURE 2.13 Muscle *a* is contracting at an angle less than 90°. Its force can be split into force *b*, which acts perpendicular to the forearm, and force *c*, which acts toward the elbow joint (note that the angle of the muscle in this figure is exaggerated for the sake of clarity).

25° from the vertical. Any angle of pull by the biceps that is not perpendicular to the forearm means that some of the force of muscle contraction is directed elsewhere and therefore plays no part in producing torque. In this example,

- *a* represents the line of action (contraction) produced by the biceps,

- *b* indicates that part of the force produces torque and lifts the forearm, and

- *c* shows that some force is directed toward the elbow joint.

Third-class levers in the human body are commonly characterized by muscles that attach near the joint and whose contraction produces tremendous torque. But this torque is transmitted along very long levers. The longer the lever, the smaller the resistance that can be moved. A force of 100 units produced by the biceps muscle is immediately reduced because the muscle pulls on the forearm at an angle other than 90°. The turning effect occurring where the muscle is attached to the forearm is further reduced by the length of the forearm. By the time it reaches the athlete's hand, it has been reduced considerably. But an athlete does gain from this arrangement.

Applied Sports Anatomy

So what do we gain from this arrangement? Think of an athlete with long arms. Long arms mean

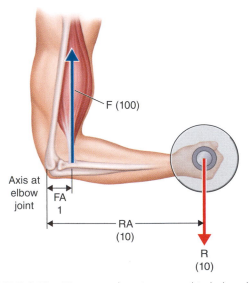

FIGURE 2.12 Biceps curl acting as a third-class lever system from within the human body.

APPLICATION TO SPORT

Lever Manipulation in Sport

The axis of rotation and the associated lever are the fundamental components for rotation. By changing our body position in sport, we can manipulate the axis to change the type of lever system in play, but the predominant lever system in the human body is a third-class lever. This system is designed for speed rather than strength. Humans are structurally designed to move things fast, so we must consider this with regard to sport. A typical example of a rotating lever system in sport is running gait. To run, we predominantly depend on our legs, and each leg segment (thigh and shank) rotates about a joint axis, namely our hips, knees, and ankles. The lever at each of these joints is third class, meaning that the muscle attachment at the joint is very close to the axis of rotation, and the resistance to move the joint is farther away. This means that a small contraction of the muscle will result in a larger movement of the lever; thus, our limbs are designed for speed. Typically, someone with long levers, or long limbs, is more suited to speed activities like running and jumping. An athlete with relatively short levers, or shorter limbs, who therefore

Kai Schwoerer/Getty Images

doesn't have to move limbs as far, is generally better suited to strength activities like powerlifting and the shot put.

that an athlete's hand travels at great velocity over a huge distance when the arm is swung in an arc. If an athlete can generate sufficient force in the muscle, a long limb can move a light resistance (such as a ball, club, or discus) over an immense range at tremendous speed. Therefore, discus throwers in the Olympic Games are huge, long-armed athletes (Ferro et al. 2001). The more powerful the athlete's muscles are and the longer the arms are, the better the performance is. Short limbs, on the other hand, give up less force along their length than do long limbs.

You'll see what we mean by this statement in the following fantasy competition between a 1.68 m (5 ft 6 in.) male gymnast weighing 59 kg (130 lb) and a giant 2.14 m (7 ft) professional basketball player who weighs 136 kg (300 lb). Let's imagine for the sake of this scenario that these two athletes have similar strength in their pectoral and latissimus muscles. We now ask both athletes to attempt an iron cross on the rings, an extremely difficult gymnastics skill. An iron cross, as you will see in figure 2.14, requires an athlete to contract the latissimus and pectoral muscles. These muscles do most of the work in pulling the arms and hands downward on the rings and simultaneously, as part of their

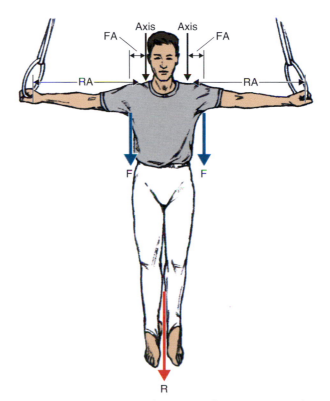

FIGURE 2.14 Force and resistance lever systems in the human body during the iron cross exercise.

contraction, in pulling the body up. This action stops the athlete's body from dropping down out of the cross position. What are the key features in this scenario?

- A basketball player with the physique that we've described is immediately at a disadvantage in this skill because long arms are long levers and, in a mechanical sense, form huge resistance arms.
- The resistance is the basketball player's body weight pulling downward.
- The axis of rotation is at his shoulder joint, and his pectoral and latissimus muscles, which provide force, have to work with a short force arm (see figure 2.8).

Of course, because they are human, both basketball player and gymnast suffer when trying to complete the fantasy iron cross. But for our basketball player, the force produced by muscle contraction in the pectoral and latissimus muscles, multiplied by a small force arm, faces off against a resistance of 136 kg (300 lb) of body weight multiplied by a huge resistance arm—the enormous length of the basketball player's arms. For our basketball player to hold an iron cross and not drop down toward the floor, the turning effect (torque) produced by force multiplied by force arm must equal the torque produced by the resistance multiplied by the resistance arm. A gymnast with shorter arms has a much better chance of success in this skill. Short arms in a mechanical sense mean short levers and small resistance arms. The torque produced by the gymnast's lighter body weight multiplied by these short resistance arms is considerably less than that of a heavyweight basketball player with exceptionally long arms.

Let's consider the issue of body weight for a moment. Our basketball player has 136 kg of body weight acting as the resistance. Even though his pectoral and latissimus muscles collectively battle 136 kg of body weight pulling downward, an iron cross would demand strength in these muscles that you'd find only in a bionic man. The contractile force that the basketball player's muscles would have to produce for an iron cross could even tear the muscles away from where they attach to the bones!

The basketball player's muscles get little help from the short force arm they must work with, and they fight immense resistance (his body weight) multiplied by a huge resistance arm (the length of his arms). These characteristics are reason enough why athletes with the physique of basketball play-

ers seldom venture into the sport of gymnastics. This somatotype (the human body's individual shape) often forms the natural selection of humans into that particular sport. For example, most male gymnasts are around 1.68 m (5 ft 6 in.) tall and weigh between 54 and 59 kg (119 and 130 lb).

- Females are even smaller, averaging 1.52 m (5 ft) and weighing no more than 49.9 kg (110 lb).
- Gymnasts benefit not only from having short arms but also from having minimal body weight. Their body weight is often less than half that of basketball players in the National Basketball Association.

Success in an iron cross (and other difficult gymnastics skills) is much more likely for those built like a gymnast. Elite gymnasts are super strong, have short limbs, and diet to control their body weight. A little extra weight (i.e., a kilogram or a few pounds) can mean the difference between success and failure in many gymnastics skills.

Now let's take our basketball player and gymnast onto the track to compete against each other in a discus competition. When they throw a discus, we find that a taller, heavier, long-limbed athlete has a distinct advantage. A spin across the ring by our long-armed basketball player gives the discus incredible velocity at the instant of release because it is pulled around a circular pathway made enormous by the length of his arms. We will look at this in more detail in chapter 7, Angular Kinetics in Sport. Proof of the importance of limb length for a discus thrower was demonstrated in the 2016

AT A GLANCE

Human Lever Systems

- The components of a lever system are the axis (or fulcrum), the force arm, and the resistance arm.
- On a first-class lever, A (the axis) is positioned between the resistance and the force. On a second-class lever, R (the resistance) is positioned between the axis and the force. On a third class-lever, F (the force) is positioned between the axis and the resistance.
- Third-class levers are the most common in the human body. A muscle attachment is near the axis, and when the muscle contracts the torque is transmitted along the long lever of the bone.

Rio de Janeiro Olympics; the men's discus gold medalists stood 2.07 m (6 ft 9 in.) tall and weighed 120 kg (265 lb).

Our gymnast loses in the discus event. If both a gymnast and a basketball player were to spin across the ring at the same rate, a discus in the hand of the long-armed basketball player would travel faster. What could our gymnast do to compensate?

- He could spin at such a phenomenal rate that he's a blur as he goes across the ring. But this feat would be difficult to accomplish, and few small athletes compete seriously in the discus event these days.

- Long arms are a great advantage in an event like the discus and a disadvantage in most gymnastics skills. All this depends on what is required by the sport. In some sport skills being heavy and having long arms and legs is better; in other sports these characteristics become a disadvantage.

Finally, coming back to our bicycle pedal example, it is worth mentioning that when the cyclist changes gears, the different diameters of the gears cause a change in length of the resistance arm, which then makes pedaling easier or harder. This adjustment allows the type of rotational movement we are trying to create in sport. We can manipulate our lever system by changing the force applied as well by modifying the lever arm, an ability that is useful in sport mechanics.

SUMMARY

- Levers are simple machines that incorporate a rigid object that rocks or rotates around an axis, or fulcrum. Force is applied at one position on the lever, and a resistance applies its own force at another.

- The two most important functions of a lever system are the ability to vary the force or alter the speed and distance. Both cannot occur at the same time.

- There are three classes of levers. In a first-class lever, the axis is positioned between the force and the resistance. First-class levers can be made to magnify either force or speed and distance. In a second-class lever, the resistance is positioned between the axis and the force. Second-class levers magnify force at the expense of speed and distance. In a third-class lever, the force is positioned between the axis and the resistance. Third-class levers magnify speed and distance at the expense of force.

- Third-class levers predominate in the human body. Most muscles in the human body apply great force to move light resistances over large distances at great speed.

- Levers produce a turning effect called torque. Torque is increased by magnifying the applied force or the distance from the axis of rotation over which the force is applied (or both).

- Levers are critical in sport performance, and with simple manipulations such as changing the grip the lever length can be modified.

KEY TERMS

antagonist	mass
agonist	resistance
axis of rotation	resistance arm
first-class lever	second-class lever
force arm	third-class lever
fulcrum	torque
lever	weight

REFERENCES

Ferro, A., A. Rivera, and I. Pagola. 2001. "Biomechanical Analysis of 7th World Championships in Athletics Seville 1999. 400 Metres." *New Studies in Athletics* 16 (2): 52-60.

Kageyama, M., T. Sugiyama, Y. Takai, H. Kanehisa, and A. Maeda. 2014. "Kinematic and Kinetic Profiles of Trunk and Lower Limbs During Baseball Pitching in Collegiate Pitchers." *Journal of Sports Science and Medicine* 13 (4): 742-50.

McKean, M.R., P.K. Dunn, and B.J. Burkett. 2010. "The Lumbar and Sacrum Movement Pattern During the Back Squat Exercise." *Journal of Strength and Conditioning Research* 24 (10): 2731-41.

Nesbit, S.M., and R. McGinnis. 2009. "Kinematic Analyses of the Golf Swing Hub Path and Its Role in Golfer/Club Kinetic Transfers." *Journal of Sports Science and Medicine* 8:235-46.

Tortora, G., B. Derrickson, B. Burkett, D. Dye, J. Cooke, T. Diversi, M. Mckean, R. Mellifont, L. Samalia, and G. Peoples. 2016. *Principles of Anatomy and Physiology*, 1st Asia-Pacific edition. Milton, Queensland, Australia: John Wiley & Sons Australia.

www Visit the web resource for review questions and practical activities for the chapter.

3

Sport Mechanics Fundamentals

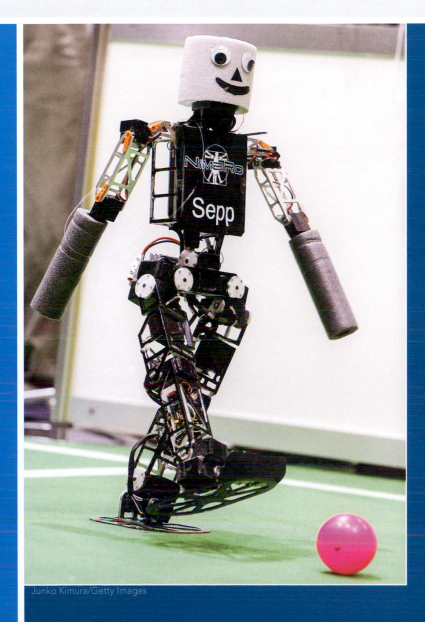

━━ **When you finish reading this chapter, you should be able to explain** ━━

- human motion;
- the mechanical terms
  - *kinematics* and *kinetics*;
  - *linear motion* and *angular motion*;
  - *statics* and *dynamics*;
- the application of sport mechanics to resistance training; and
- measurement and evaluation in sport mechanics.

This chapter lays the foundations to describe human motion, that is, what is happening when a human performs exercise or sport. The core features in illustrating sport mechanics are the broad movement classifications of kinematics and kinetics. To keep with theme of the applied text, the application of these fundamentals is presented in the measurement section at the end of this chapter.

How to Describe Human Motion

The study of sport mechanics embraces many disciplines or terms. In the early 1970s the most commonly accepted term to describe human motion was the discipline of **biomechanics**, a combination of *bio*, the study of human biological systems, and *mechanics*, the study of mechanical principles. In the previous chapter we investigated the anatomical aspects of sport mechanics. In this chapter we take those standard reference systems and explore the mechanical aspects of human movement. The key to describing the mechanical aspects, in an anatomical reference system, is the specialty of kinematics and kinetics.

Kinematics and Kinetics

Kinematics describes motion—how something is moving. This meaning can include things like the sequence of movement for various body segments or the measurement of key variables of distance, velocity, and acceleration.

Kinetics describes the forces that create motion or are generated from moving. This meaning can include things like ground reaction force, friction, torque, and internal muscle forces. Another way to consider these two definitions is that kinematics typically describes human movement, whereas

kinetics attempts to determine the cause (or effect) of human motion.

Linear and Angular Motion

Both kinematics and kinetics can be used to describe and define human motion in linear or angular formats. The movement of an object can be classified in three ways:

- Linear (in a straight line)
- Angular (in a circular or rotary fashion)
- A mix of linear and angular, which we simply call **general motion**

In sport, a mix of linear and angular movement is most common. Angular movement plays the dominant role because most of an athlete's movements result from the swinging, turning action of the athlete's limbs as they rotate around the joints (Bennett et al. 2009).

Linear motion describes a situation in which movement occurs in a straight line. Linear motion can also be called **translation**, but only if all parts of the object or the athlete move

- the same distance,
- in the same direction, and
- in the same period.

As you can imagine, translation rarely occurs in an athlete's movement because some parts of an athlete's body can be moving faster than other parts and not always exactly in the same direction. For example, an athlete in the 100 m sprint wants to travel the shortest distance from the start to the finish. The shortest distance is a straight line. Yet sprinting is produced by a rotary motion of the limbs as they pivot at the athlete's joints, and the athlete's center of gravity rises and falls during each stride.

Many terms are used to refer to **angular motion**. Coaches talk of athletes rotating, spinning, swinging, circling, turning, rolling, pirouetting, somersaulting, and twisting. All these terms indicate that an object or an athlete is turning through an angle, or number of degrees. In sports such as gymnastics, skateboarding, basketball, diving, figure skating, and ballet, the movements used by athletes include quarter turns (90°); half turns (180°); and full turns, or "revs" (revolutions), which are multiples of 360°. Slam-dunk competitions are a great example of basketball players showing off their "360s."

To produce angular motion, movement has to occur around an axis. You can think of an axis as the axle of a wheel or the hinge on a door. An athlete's body has many joints, and they all act as axes. The most visible rotary motion occurs in the arms and legs. The upper arm rotates at the shoulder joint, the lower arm at the elbow joint, and the hand at the wrist. The hip joint acts as an axis for the leg, the knee for the lower leg, and the ankle for the foot. Movements like walking and running depend on the rotary motion of each segment (e.g., foot, lower leg, and thigh) of an athlete's limbs as they rotate around the joints.

Because this angular rotation is occurring about our anatomical hinges (or joints), the length of our limbs will influence how effectively we can produce the desired movement. Because the length of our limbs varies and changes as we mature, this variation in size can lead to a variation in the scale (or size) of the sports equipment. The most common example is the difference between an appropriate child-size sports equipment and a professional athlete's sports equipment (Timmerman et al. 2015).

All human motion is best described as general motion, a combination of linear and angular motion. Even those sport skills that require an athlete to hold a set position involve various amounts of linear and angular motion.

- A gymnast balancing on a beam and a ski jumper crouching in an aerodynamic position during the acceleration before takeoff are good examples. In maintaining balance on the beam, the gymnast still moves, however slightly. This movement may contain some linear motion but will be made up primarily of angular motion occurring around the axes of the gymnast's joints and the location where the gymnast's feet contact the beam.

- The ski jumper holding a crouched position attempts to reduce air resistance to a min-

---

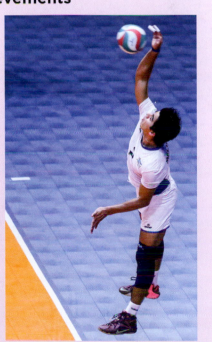

FIGURE 3.1   A wheelchair athlete exhibits a combination of angular and linear motion.

Friedemann Vogel/Getty Images

imum and accelerate as much as possible before takeoff. Sliding down the in-run holding a crouched position is a good example of linear motion. But the athlete never fully maintains the same body position throughout, and the in-run is not straight throughout, so any motion that the ski jumper makes will be angular in character.

Perhaps the most visible combination of angular and linear motion occurs in a wheelchair race. The swinging, repetitive angular motion of the athlete's arms rotates the wheels. The motion of the wheels carries both the athlete and the chair along the track. Down the straightaway, the athlete and chair can be moving in a linear fashion (Tolfrey et al. 2012). At the same time the wheels and the athlete's arms exhibit angular motion (see figure 3.1). This combination of angular and linear motion is an example of general motion.

Statics and Dynamics

The final aspect to describe human motion is the area of **statics** and **dynamics**. These are also typical subdivisions of mechanics. Just like kinematics and kinetics, or linear and angular motion, human motion may or may not fall exactly into one of these categories, but we need to use these delineators to observe and quantify human movement.

- Statics describes motion that is in a constant state of motion, which can include moving at a constant velocity or not moving at all when at rest (because you are in a constant state of no movement).

- Dynamics describes motion that is undergoing change, that is varying. Dynamics can include things like increasing your velocity. Note that any change in motion is subject to acceleration (or deceleration).

AT A GLANCE

Describing Human Motion

- Kinematics describes motion, or how something is moving.
- Kinetics describes the forces that create motion or are generated from moving.
- Movement can be linear (in a straight line), angular (in a circular or rotary fashion), or a mix of linear and angular.
- Statics describes motion that is in a constant state of motion.
- Dynamics describes motion that is undergoing change, that is varying.

To keep in line with the applied nature of this textbook, all these mechanical descriptors (kinematics and kinetics, linear and angular, statics and dynamics) can be put into practice in areas such as resistance-training devices for athletes.

Applying Sport Mechanics to Resistance Training

One of the common mechanical features of sport training is the development of a **resistance-training** program, which uses some form of mechanical advantage (or disadvantage) to provide a load (or resistance) to the athlete. When athletes perform these exercises, they overload the muscular skeletal system. If the optimal amount of overload is provided, the human body will repair stronger and faster, which is a desired outcome for any athlete. To ensure the safety of the athlete using resistance training it is imperative they use a **spotter** to provide assistance if needed.

By using our mechanical knowledge of the various forms of resistance, we can generate a range of resistance exercises. Within these various forms of exercise, we can also vary the type of mechanical load. For example, the athlete's training session may focus on eccentric movement in which the athlete moves the weight at a velocity that causes the muscle to lengthen under load.

Consider a biceps curl. You start with your elbow fully extended when viewed from the sagittal plane and then proceed to curl your elbow and raise the weight. On the way up the anatomical muscle of the biceps is contracting (getting shorter) as you lift the weight. Then to return to the starting position, on the way down the biceps muscle needs to extend (get longer) (see figure 3.2).

- Depending on the velocity of movement you can change the **forces** that are applied to the biceps; that is, you can generate concentric forces by having the muscle contract under load.
- Alternatively, you can generate eccentric forces by having the muscle extend while under load. This action requires a slower, controlled technique.

Each of these causes a different force on the muscular skeletal system and therefore a different training response. How big an area this force is applied over will influence the **pressure** that is associated with this force. The mechanical resistance devices can be categorized in the following areas:

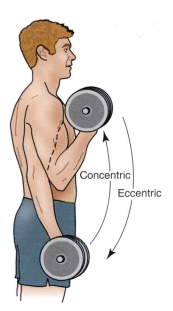

FIGURE 3.2 Biceps curl with concentric and eccentric movements.

- Machine-weight exercises
- Free weights
- Isometric resistance
- Variable machine devices
- Nonweight resistance devices

Machine-Weight Exercises

When an athlete starts a resistance-training program she most likely will not have developed the proper and safe movement pattern, or technique. Therefore, the common starting point for a **strength** and conditioning program is to perform exercises using a weight machine, sometimes known as pin-weight machine (figure 3.3).

In this setup the equipment usually has some form of resistance, say metal plates, which is attached by a pulley system to an attachment handle. The amount of resistance in the machine-weight device can vary, depending on the orientation of the pulleys. In theory, if the selected weight is, say, 20 kg (44 lb), then the athlete will be applying a constant mechanical resistance when she exercises.

As a machine exercise, this movement would have a form of guides or rails to keep the weights moving in a defined path. More important, if the athlete is starting to fatigue or her technique is variable, she may drift away from the safe movement path, which could cause injury. Naturally, the inclusion of a guide or track to move along will reduce the work required of the stabilizing

FIGURE 3.3   Typical weight machine.

Glow Wellness/Getty Images

muscles to ensure a safe and effective movement. In essence, the athlete just has to push or pull the weight along the track.

Free Weights

As the athlete progresses through her training program and her technique for exercise becomes more efficient, she typically advances to a free-weight exercise. The typical example is the use of free weights such as a barbell or dumbbell. By selecting, say, a 20 kg (44 lb) barbell, the athlete will be applying a constant mechanical resistance when she exercises. The term *constant* is flagged because although the athlete is using the same weight, the constantly changing lever arm within the human body means that the resistance varies. Because the weight is not connected to a machine or guide, the athlete is free to move the weight in almost any direction, depending on the desired loading effect (figure 3.4). As the name suggests, this training format provides a constant amount of resistance to the athlete (Walsh et al. 2007).

A key component of the free-weight exercise is that the resistance is generated by the force of gravity. We use our knowledge of kinetics and the study of forces to apply the influence of gravity to sport mechanics. Because gravity acts only to the center of the earth, the athlete's orientation of the free weight can cause variation in the force of gravity.

Namely, if the athlete is lifting up against gravity, she will need to generate forces within her muscles to overcome the free weight. Alternatively, if she is lifting a weight that is moving down with gravity, she will need to control the movement and prepare the stop at the end of the range of movement.

Although a free weight will have a theoretical constant amount of resistance, because of the

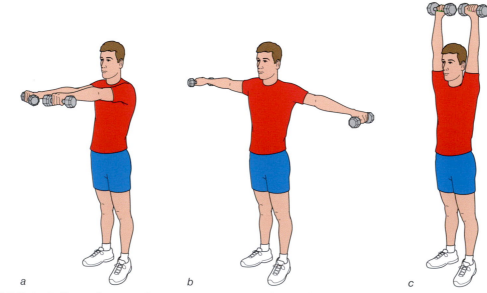

a   *b*   *c*

FIGURE 3.4   Different free-weight orientations to vary the load.

constant gravitation forces, by varying the direction of moment and the orientation of movement (with respect to gravity), the free-weight "constant" resistance can be manipulated.

Isometric Resistance

For either the machine-weight or free-weight setup, the athlete would typically perform the exercise through the full range of movement. She would start at one point of the maximum anatomical range of movement and then use her musculoskeletal system to move through the anatomical range of movement.

One variation to this type of exercise is to stop and hold the resistance (holding a weight or just the body limb segment) at a specific angle or anatomical position. For example, the athlete could move the weight until the human segment is horizontal to the ground and then hold the weight at the position for the desired length of time.

Another variation is to get into the desired anatomical position and generate the maximum amount of force while in that orientation. Because this anatomical position can be reliably repeated (i.e., the body segment kept horizontal to the ground in the same position), it provides a key foundation for accurate athlete testing or research (figure 3.5). We will expand further on measurement within sport mechanics in the next section.

FIGURE 3.5   Isometric exercise, leg squat with knee at 90° bend.

Variable Machine Devices

The standard machine resistance-training device will have a load (usually a metal plate) that is connected by a series of cables to the attachment handle. The standard pulleys are circular; they have the same radius distance from the center of the pulley to the point where the cable passes.

APPLICATION TO SPORT

Use of Cams on Weight-Training Machines

Many types of weight-training machines make use of oddly shaped cams, which are basically pulley wheels with off-center axes. They have a profile (i.e., shape) that depends on the specific exercise for which they were designed. Joined to the circumference of the cam is a cable or chain that is often attached to the weight stack. As the athlete uses the machine, the cam is forced to rotate. The rotation of the cam varies the length of the radius from the cam's axis to where the cable contacts its circumference. This variation in radius changes the resistive torque produced by the weight machine so that the athlete has to work harder at some joint angles than at others. This type of accommodating resistance takes into account the fact that the human body is stronger and more efficient at certain joint angles than at others. Generally, a cam machine is considered superior because the resistance changes with position, which is what happens in sport.

From your knowledge of anatomy and lever systems, you know that the most common lever in the human body is the third-class lever. So taking the example of the biceps curl, as shown in figure 3.2, as the athlete lifts the weight through the range of the biceps curl, the lever distance (the perpendicular distance from the axis of rotation to the weight) changes. As this distance changes so should the total amount of effort required to move the weight. In a typical machine-weight setup, the lever arm is the same (because the machine has a circular pulley with the same radius or lever arm). To address this issue, some machines incorporate a variable pulley setup, or camber (see the Application to Sport sidebar). The camber changes the lever arm distance and therefore changes the amount of effort required to move the weight. This setup better reflects an athlete's "real" resistance when moving body segments (Bennett et al. 2009).

Nonweight Resistance Devices

The final type of resistance training devices does not rely on the mass of the weight and the forces of gravity to provide the resistance. These devices use other mechanical forms to generate resistance. The common formats are hydraulics, pneumatics or air resistance, and manual brakes (figure 3.6). For each of these scenarios the machine engages the resistance as the athlete exercises.

- A hydraulic exercise machine capitalizes on the mechanical setup for a hydraulic ram. In this setup as the user pumps or tries to move the hydraulic fluid through the chamber, the size of the internal openings are reduced to create resistance to movement.

- A pneumatic exercise machine follows principles similar to those used in the hydraulic ram, but uses air instead of hydraulic oil to provide the resistance. Depending on this setup these machines can sometimes generate less resistance than oil, but if they break or leak, only air escapes, rather than an oily mess.

- A manual brake machine uses, as the name suggests, a form of clutch or brake to provide resistance as the athlete exercises. The amount of resistance, or braking, may be variable.

FIGURE 3.6   Hydraulic exercise machine.
Courtesy of AeroStrength Hydraulic Fitness Equipment.

Measurement and Evaluation in Sport Mechanics

This final section combines each of the areas of describing human motion and the application of sport mechanics to resistance exercise with the important outcome—measurement. How do we validly and reliably measure kinematics, kinetics, or the influence of resistance (Fulton et al. 2009)? This section explores the key requirements for measurement in sport mechanics, which can be applied in the subsequent discipline-specific chapters.

With current-day technology our world is full of "devices," "tools," and "apps" that *can* measure what is happening in sport mechanics. The trap is in blindly believing that the number or quality generated by the electronic device is valid, reliable, and relevant. As students of sport mechanics we need to embrace new technology, because in many instances it can provide us with new knowledge, but we need to be confident that the measurement is correct.

Why Measure or Evaluate in Sport?

As flagged in chapter 1, the key delineator in describing and measuring sport mechanics is the format of the desired output. Do you require a **quantitative** output that describes numbers or a **qualitative** output that describes the quality of the performance?

APPLICATION TO SPORT

Measuring Makes a Difference

To understand what is happening in sport or to answer the question of how something was accomplished, the fundamental requirement is to have an accurate and reliable measure. Almost every high-performance athletic program regularly conducts an athletic performance measure, such as vertical jump height for a volleyball player, reaction time for a tennis player, or maximum bench press for a football player. These measures can provide valuable feedback about whether the athlete is improving, thus indicating the effectiveness of the current training program. The knowledge from conducting these regular measures can also monitor the progress of injury rehabilitation. For example, when the player reaches his previous scores in training, he should be ready to come back into the competition proper. Finally, regular measures can provide a great screening tool for talent identification, particularly when scouting or recruiting new players. The key is to have confidence in the measures, which is achieved when the measures are valid and reliable.

Another way to differentiate these processes is to classify a quantitative approach as measurement in the sport and a qualitative approach as evaluation of the sport. Despite the differences in the format, either output relies on the underlying measurement being valid and reliable (Clark et al. 2009). You might make measurement in sport for a number of reasons.

- The most common reason for taking measurement in sport is to describe the current performance level (or specific aspects of the performance). This information can provide valuable feedback to the coach and sport science staff on the effectiveness of the training program. Is this approach working? Are improvements in performance being made?

- A variation on the description of performance is to focus on an outstanding elite performance and try to understand how it was achieved. These measures can then define the future aspiration for a top performance in that sport.

- Measurement in sport can also identify why a sport-related injury occurred or why a certain type of injury is becoming common.

- Relevant and reliable measures in sport can lead to a prediction of athletic potential. Modeling of these measures can provide some framework to what the future may hold.

Accuracy of Measurement

Because this text is focused on sport mechanics, which is predominately a quantitative process, we will concentrate more on the accuracy of quantitative measurement. This knowledge on the effectiveness of the performance can certainly be used to describe the quality or evaluation of the sporting performance. Two core questions define the accuracy of measurement:

- Does the measurement measure what the intended variable relates to? This question addresses the **validity** of the measurement.

- Can the measurement be repeated and provide a consistent value? This question addresses the **reliability** of the measurement.

These descriptors for the measure are independent of each other; that is, a measurement may be reliable and but not valid, or valid but not reliable.

As the name suggests, a valid measure needs to measure the variable you are concerned with. An example in sport mechanics would be the validity of a measurement of jumping height. The common ways to measure height would be with a ruler, tape measure, or stadiometer. To ensure that any of these methods is correct, it would need to be compared with a calibrated or industry-standard device. That calibration would confirm that the measured height of, say, 1.2 m (3.9 ft) is correct.

For the measure to be reliable the instrument (or device) needs to produce the same measure every time. An example in sport mechanics would be a set of scales used to measure weight. The scales need to measure the athlete's weight reliably every time. If the scales are consistently off by, say, 3 kg (6.6 lb), they may be producing a reliable measure, but the measure is inaccurate and therefore not valid.

AT A GLANCE

Measuring Sport Mechanics

- The most common reason for taking measurement in sport is to describe the current performance level. This information can also lead to a prediction of athletic potential.

- A valid measure is one that measures what the intended variable relates to.

- A reliable measure can be repeated and provides a consistent value.

SUMMARY

- Kinematics describes motion, or how something is moving.
- Kinetics describes the forces that create motion or are generated from moving.
- In sport, a mix of linear and angular movement is common.
- Linear motion describes a situation in which movement occurs in a straight line.
- Statics describes motion that is in a constant state of motion.
- Dynamics describes motion that is undergoing change, motion that is varying.
- One of the common mechanical features of sport training is the development of a resistance-training program.
- The common starting point for a strength and conditioning program is to perform exercises using a weight machine, sometimes known as a pin-weight machine.
- When using a free weight the athlete is free to move the weight in almost any direction, depending on the desired loading affect.
- An isometric exercise requires the athlete to stop and hold the resistance at a specific angle or anatomical position.
- With a variable machine exercise the camber changes the lever arm distance and therefore changes the amount of effort needed to move the weight.
- Some machines can use other mechanical forms to generate resistance. The common formats are hydraulics, pneumatics or air resistance, and manual brakes.
- Any measurement needs to be valid; that is, the measurement must measure what the intended variable is.
- A reliable measure is one that can be repeated and provides a consistent value.

KEY TERMS

angular motion	quantitative
biomechanics	qualitative
dynamics	reliability
force	resistance training
general motion	strength
kinematics	spotter
kinetics	statics
linear motion	translation
pressure	validity

REFERENCES

Bennett, J., M. Sayers, and B. Burkett. 2009. "The Impact of Lower Extremity Mass and Inertia Manipulation on Sprint Kinematics." *Journal of Strength and Conditioning Research* 23 (9): 2542-47.

Clark, R., A. Bryant, Y. Pua, P. McCrory, K. Bennell, and M. Hunt. 2009. "Validity and Reliability of the Nintendo Wii Balance Board for Assessment of Standing Balance." *Gait and Posture* 31 (3): 307-10.

Fulton, S. K., D. B. Pyne, and B. Burkett. 2009. "Validity and Reliability of Kick Count and Rate in Freestyle Using Inertial Sensor Technology." *Journal of Sport Sciences* 27 (10): 1051-58.

Timmerman, E., J. De Water, K. Kachel, M. Reid, D. Farrow, and G. Savelsbergh. 2015. "The Effect of Equipment Scaling on Children's Sport Performance: The Case for Tennis." *Journal of Sports Sciences* 33 (10): 1093-1100.

Tolfrey, V., B. Mason, and B. Burkett. 2012. "The Role of the Velocometer as an Innovative Tool for Paralympic Coaches to Understand Wheelchair Sporting Training and Interventions to Help Optimise Performance." *Sports Technology* 5 (1-2): 20-28.

Walsh, J., J. Quinlan, R. Stapleton, D. FitzPatrick, and D. McCormack. 2007. "Three-Dimensional Motion Analysis of the Lumbar Spine During 'Free Squat' Weight Lift Training." *American Journal of Sports Medicine* 35 (6): 927-32.

www Visit the web resource for review questions and practical activities for the chapter.

4

Linear Motion in Sport

When you finish reading this chapter, you should be able to explain

- factors that influence an object's or athlete's acceleration;
- methods of measuring linear motion (speed, velocity, and acceleration);
- projectile motion; and
- factors that influence projectile motion

Now that we have covered the fundamentals of anatomy and sport mechanics, we can go into more detail on each of the core components that make up applied sport mechanics. We begin with constant motion, or **linear motion**. The mechanical principles we discuss in this chapter put into practice those we've already talked about in chapters 2 and 3. As you develop your knowledge about mechanics, you'll notice how all these principles tie in to one another. Just as important, you'll see that every action that athletes make in a sport skill involves several mechanical principles that occur simultaneously. Good technique in sport is based on making the best use of these mechanical principles. So let's see what happens when athletes apply force with their muscles (the anatomical foundations) and put themselves, and objects such as bats and balls, in motion (the mechanical foundations). As in previous chapters we begin with the standard international terminology to describe linear motion in sport.

Linear Motion Measures (Speed, Velocity, and Acceleration)

Just as the terms *mass* and *weight* are used interchangeably (sometimes incorrectly), a similar situation occurs with ***speed*** and ***velocity***. Although both terms indicate how fast an object is traveling with respect to time, the definitions differ slightly. Speed is a **scalar** measure indicating how fast an object is traveling, measured by dividing the length or distance traveled by the time, but speed does not quantify the direction of travel. Velocity, on the other hand, is the change in position divided by the time.

If an elite sprinter runs 100 m in 10 s, we know that the athlete has run a certain distance (100 m, or 109.4 yd) in a certain time (10 s). From this information you can work out the sprinter's average speed, which is 100 m divided by 10 s = 10 m/s (10.9 yd/s), or 36 km/h (22.4 mph).

When running 100 m on a straight track, because the direction of travel is in a straight and consistent line, the change in position is also 100 m, so the calculation for speed and velocity is the same in this instance. Sometimes, however, we need to know in which direction, as well as how fast, the object is traveling (i.e., north or south, or positive or negative). In these situations, velocity is the better term to use. For example, when kicking a ball, as the ball takes off we can look at how fast the ball is traveling:

- We measure how fast the ball is traveling in the horizontal direction (that is, with respect to the horizontal ground).
- We measure how fast the ball is traveling in the vertical direction (that is, with respect to the vertical).
- Or, to measure how fast just the ball itself is traveling, we measure the **resultant** of these two components (that is, the resultant of the horizontal and vertical components).
- To measure how fast the ball travels in these planes, we measure velocity, not speed.

The velocity that the sprinter averaged over a distance of 100 m is 36 km/h (22.4 mph)—nothing more. These numbers don't tell you the sprinter's top velocity, which could be as high as 42 km/h (26 mph), and they don't tell you anything about the sprinter's **acceleration** or **deceleration**, which is the rate at which velocity (or speed) changes (Lee et al. 2010).

A sprinter who averages 36 km/h (22.4 mph) over 100 m runs faster and slower than 36 km/h during different phases of the race. Why? Because immediately after the starter's gun goes off, the athlete is gaining velocity and for a while runs much slower than 36 km/h. The athlete then has to run faster somewhere else in the race to produce an overall average 36 km/h over the whole distance.

Rates of acceleration vary dramatically from one athlete to another. Some athletes rocket out of the blocks and have tremendous acceleration over the first 40 m of a 100 m race. Thereafter their rate of acceleration drops off, and as they get close to the finishing tape they may even decelerate. Athletes who raced against multiple Olympic champion Carl Lewis were well aware that he could still be accelerating at the 70 m mark in the 100 m dash.

- Carl Lewis' running technique incorporated a rate of acceleration that may have been less than that of his opponents at the start of the race, but his acceleration continued longer.

- Over the last 30 m, Lewis frequently caught and passed athletes who were "tying up" (i.e., breaking proper form because of fatigue) and subsequently decelerating.

In the 400 m event, the sport mechanical analysis breaks the race into 50 m chunks to allow a more specific average velocity measure. The velocity profile for the Olympic champion and world-record holder Michael Johnson, who broke the 400 m world record in the time of 43.18 s in 1999, is shown in figure 4.1 (note that this world-record time was not broken until the 2016 Rio de Janeiro Olympic Games). Johnson's maximum velocity was at the 150 m mark, and the key difference between Johnson and his opponents was the smaller amount of drop-off after each 50 m interval.

Athletes can reduce their rate of acceleration and still increase velocity. As long as acceleration exists, even if it is minimal, velocity will increase (because acceleration is the rate of change of velocity). If deceleration occurs, velocity will naturally decrease. How much an athlete's velocity increases or decreases depends on the rate of acceleration or deceleration.

Uniform acceleration and **uniform deceleration** mean that an athlete, or an object, speeds up or slows down at a regular rate. An example of uniform acceleration occurs when a four-man bobsled slides down the track in the Winter Olympics and accelerates to a speed of 4.6 m/s (15 ft/s) by the first second, 9.1 m/s (30 ft/s) by the second, and 13.7 m/s (45 ft/s) by the third.

- For every second that the bobsled is moving, it is increasing speed at a uniform rate of 4.6 m/s (15 ft/s).

- You write this acceleration as 4.6 m/s/s or 4.6 m/s^2 (15 ft/s/s or 15 ft/s^2).

- Notice that you use one distance unit (i.e., 4.6 m) and two time units (i.e., s/s) whenever you refer to acceleration. This indicates the rate of change of velocity, or the amount of velocity added (i.e., 4.6 m/s), with each successive time unit (i.e., 1 s) that passes.

- If the bobsled decelerates at a uniform rate, then the reverse occurs. In this case it is slowing, or losing velocity, at a uniform rate.

Uniform acceleration and deceleration do not often happen in sport. When athletes (or objects such as balls or javelins) are on the move, varying oppositional forces, ranging from opponents to air resistance, cause their acceleration (or deceleration) to vary or, in other words, to be nonuniform. One of the best examples of uniform acceleration and deceleration occurs in flights of short duration such as in high jump, long jump, diving, trampoline, and gymnastics.

In these situations, air resistance is so minimal that it can be considered negligible. Gravity uniformly slows, or decelerates, the athletes as they rise in flight by a speed of 9.8 m/s (32 ft/s) for every 1 s of flight (i.e., 9.8 m/s^2) and then accelerates them at a uniform rate of 9.8 m/s^2 on the way down. Sometimes you'll see deceleration described as **negative acceleration** and acceleration as **positive acceleration**. A minus sign in front of 9.8 m/s^2 (i.e., −9.8 m/s^2) indicates that the diver is decelerating at a rate of 9.8 m/s^2 for each second that he is rising in the air.

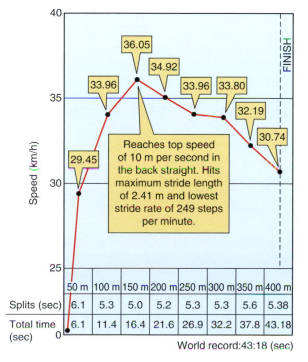

FIGURE 4.1 The velocity profile of Michael Johnson's 400 m world record.

How to Measure Linear Motion (Speed, Velocity, and Acceleration)

Because velocity is simply the change in position divided by the change in time, all that is required to calculate linear motion is to measure the distance and time variables for a linear movement (say, a 100 m sprint). The simple tools are a tape measure and stopwatch. We just measure the distance that has been traveled and the time to travel to a specified position, and then calculate the velocity. For multiple measures, say every 10 m in the 100 m event, these quantities can easily be set up in a spreadsheet because the calculations are repetitive, as shown in the sample spreadsheet in table 4.1.

- In the first column we set up the predefined distance intervals, such as every 10 m.

- In the second column we measure the time the athlete passes through this distance interval.

- From the information in the first two columns, we calculate the variables in the subsequent columns. The third column is the change in time to run that 10 m interval, which is the split time for that segment.

- The fourth column, velocity, is the distance (10 m) divided by the split time to run that interval.

- The fifth column is the difference in velocity between the interval and the previous one.

- The sixth column, acceleration, is the change in velocity divided by the split time.

We can measure the velocity of the whole body (for example, the whole body of a runner), or we can measure the velocity of individual components (for example, the velocity of just the runner's leg). Individual parts of the runner's body will move at different rates over a distance of 100 m. For example, the area around the runner's belly button (commonly the location of the center of gravity) would move at a pretty constant rate, yet segments like each arm (or leg) can move faster (and slower) within each step. So we can measure the velocity of the whole body and assume that it acts as a rigid body, or we can measure the individual components such as the leg or even the thigh (of course, if we wish, we can do both). We'll discuss the different ways to measure gravity in the section on gravity.

Using Technology to Measure Linear Velocity

The capability of current-day technologies are constantly expanding, particularly in the electronic and Internet arena. In an applied sense, sport mechanics can adopt and capitalize on this sector by incorporating these technologies to automate (or semiautomate) the process of measuring **linear velocity**. One of these technologies is timing gates—electronic devices that use a laser or infrared beam and a reflective marker (Barris and Button 2008). As the runner breaks the beam, the device triggers the internal stopwatch to record the runner's time. Using a tape measure, you can accurately position the timing gates so that they are situated at the correct distance. The runner's passage through each set of gates will trigger the stopwatch to measure the elapsed time. The internal computer of the timing gates then uses a setup similar to the spreadsheet you've just seen to calculate the change in distance divided by the change in time to determine the speed or velocity.

Other examples of technology that can measure velocity are smartphones and associated apps. All these devices use the same process as the timing gates. That is, they measure time, and when this measure is presented with the known distance, a measure of speed or velocity can be calculated (Stamm et al. 2013). All these devices or systems use the same fundamental process; the key difference is that these processes can be automated and

TABLE 4.1 Sample Spreadsheet to Measure Linear Motion (Velocity and Acceleration)

Distance	Time (s)	Change in time (s)	Velocity (m/s)	Change in velocity (m/s)	Acceleration (m/s^2)
10	2.05	2.05	4.87	4.87	2.38
20	3.26	1.21	8.26	3.39	2.80
30	4.42	1.16	8.62	0.36	0.31
40	5.49	1.07	9.35	0.73	0.68
50	6.61	1.12	8.93	−0.42	−0.38

Measure for the first two columns (distance and time) and from these data calculate for the remaining columns.

presented in real time, making the task of coaching or assessing a human performance easier. These types of devices are commonly used in team sport environments because the automated speed or velocity calculations eliminate the manual work required and (potentially) lessen the chance for human error.

At the upper end of technology, modern radar equipment that is used in sport research can provide information about speed and distance of an athlete or object. Radar technology can calculate the speed of a baseball as it leaves the pitcher's hand or as it leaves the bat at the instant it is hit. This equipment is now so precise that it can determine over what distance a sprinter was accelerating. It can also give the distance over which the athlete held a particular speed during a race and by how much the athlete decelerated at the end of a race.

How Linear Motion Occurs

For an object to move or remain stationary, some influence of force needs to occur; that is, a force must hold the object stationary or make it move. What exactly do we mean by force? You cannot actually see a force, but you can see and experience its effects. A **force** is a push or a pull that changes or tends to change the shape or the state of motion of an athlete or object. Here's an example to explain what we mean by *tends to*.

Imagine a weightlifter attempting to lift a barbell from the floor in a vertical direction. The athlete

---

AT A GLANCE

Linear Motion Measures

- Speed is a scalar measure indicating how fast an object is traveling, measured by dividing the length or distance traveled by the time. Speed does not quantify the direction of travel.

- Velocity is the change in position divided by the time and includes the direction of travel.

- Velocity can be in the horizontal direction, the vertical direction, or the resultant of these two components.

- Acceleration is the rate of change of velocity; an increase or decrease indicates acceleration or deceleration.

- Current-day technologies can be adopted to automate (or semiautomate) the process of measuring linear velocity.

APPLICATION TO SPORT

What Do Speed, Velocity, and Acceleration Mean in Sport?

The measure of velocity, and subsequent acceleration, is one of the most common and important measures in sport because this variable is useful for determining whether an athlete's current performance is better than a previous performance, or if it is better or worse than the opposition's. To calculate the direction and the speed at which someone or something is traveling, we need to obtain an accurate measure of the distance and time. Tests to measure distance and time, such as 30 m sprint drills, are often used in sport as part of an athlete's training. By measuring the smaller intervals within this distance, coaches can identify where an athlete is running fast and where she is slowing down. After the velocity has been calculated, the acceleration can be determined, and in sport, acceleration (the rate of change in velocity) is often what separates an athlete from an elite athlete. From this new information, the coach can devise specific training for each athlete. For example, if it is known that an athlete has a certain velocity over 30 m, technique and strength training can be developed that appropriately match this velocity.

Henry Lederer/The Image Bank/Getty Images

reaches down and pulls on the bar. If she pulls hard enough and applies sufficient force, the barbell is hoisted upward. But what happens if she doesn't apply enough force to move the barbell? In this situation you could say there is a tendency for the athlete to set the barbell in motion—it is closer to moving with the athlete pulling on it than if she wasn't pulling on it.

If another athlete adds her force in the same direction as that of the original lifter, their combined force may be sufficient to move the barbell off the ground. The tendency toward movement caused by the first athlete is turned into action by help from the second. The barbell moves. In this scenario you must assume that both athletes pull in the same direction. A different effect occurs if the second athlete pulls the barbell sideways rather than upward.

Whenever an athlete performs a sport skill, the athlete primarily produces internal force within the body by contracting the muscles, which results in external movement. The muscles contract and pull on tendons, and the tendons pull on the bones. The force produced by the athlete then competes against the external forces produced by gravity—ground reaction force, friction, air resistance, and, in many sports, the contact forces provided by opposing players. These terms will be discussed in more detail in later chapters.

Force Vectors

In the weightlifting scenario just discussed, we imagined that two lifters combined their muscular force to lift a barbell. The combination of their forces totaled a certain amount and was aimed in a particular direction. When the direction and amount of the applied force are known, the combination of these two items is called a **force vector**. The term *vector* simply means a quantity that has direction. In the case of the weightlifter, a certain amount of force was vectored in a vertical direction.

In mechanics, force vectors are often represented diagrammatically by arrows. The head of the arrow indicates in what direction the force is acting, and the length of the arrow represents the amount of force being applied. In our weightlifting example, if one athlete lifts the bar vertically and the other pulls it horizontally, the result is that the two athletes pull the barbell partially upward and partially to the side. Depending on the amount of force applied by each athlete, the barbell moves in the direction of the **resultant force vector**. The resultant force vector in this case is the equivalent

of two forces that simultaneously pull the barbell in different directions.

What occurs when two forces are applied in different directions against an object is represented in figure 4.2, which shows two athletes pulling on a large resistance. Athlete A applies 10 units of force toward the north. Athlete B applies 10 units of force toward the east. Arrow *a* represents 10 units of force, and arrow *b* represents 10 units of force. Each unit of force is given the same dimension (1 cm per unit, or, if preferred, 1 in. per unit).

A parallelogram is drawn with opposite sides and angles equal. In this particular case, each side will be 10 cm (4 in.) in length and each angle will be 90°. The diagonal *c* gives us the resultant force vector, which will be the direction of force resulting from the combined efforts of athletes A and B. Careful measurement indicates that *c* is 14.14 units in length and represents 14.14 units of force. This also means that a single force of 14.14 units (e.g., *c*) acting at 45° from the horizontal can be split into two forces—*a*, which has 10 units of force, acting vertically, and *b*, which has 10 units of force, acting horizontally. To find the outcome of several forces acting on a single object, you would need to know (1) the number of forces involved, (2) the magnitude of each individual force, and (3) the direction in which each force is applied.

Putting this combination of forces into practice, the internal vector forces within the human also are summed, resulting in the final movement pattern.

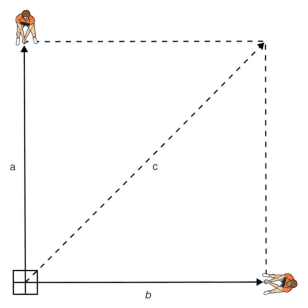

FIGURE 4.2   A parallelogram of forces. Arrow *a* shows the force applied by the athlete vertically (north); arrow *b* shows the force applied by the athlete horizontally (east); arrow *c* shows the resultant force vector.

APPLICATION TO SPORT

What Do Force and Force Vectors Mean in Sport?

A force causes or changes the state of motion of an athlete or an implement, and although we can't physically see this force, it is essential in sport. For example, when a player shoots a basketball toward the hoop, the sum of the internal musculoskeletal force vectors causes the arm to move, and the movement of the arm causes the basketball to project toward the hoop. These forces can be distributed into typical vectors, like the horizontal force that travels toward the hoop and the vertical force that moves against gravity. This understanding of the forces and force vectors is necessary to analyze the effectiveness of sport performances and to determine if and how things need to change. When you shoot the basketball, if the ball doesn't travel far enough toward the hoop, a greater horizontal force is required; if the ball doesn't travel high enough, a greater vertical force is required.

Dylan Buell/Getty Images

The human musculoskeletal system is made up of a complex arrangement of skeletal bones and muscle attachments, and when these internal forces move in a desired pattern, the result is an external movement. Depending on the orientation of the anatomical limb, these internal vector forces can create either a vertical component and a horizontal component, as shown in figure 4.3a, or a single resultant force, as shown in figure 4.3b.

Athletes often combine forces in sport to produce a desired result. Top-class soccer players know from experience how long it takes a soccer ball to travel a particular distance. They assess the speed of the forwards on their team as the forwards sprint into open field positions. When the player with the ball makes a downfield pass, several factors must be considered, including the direction and strength of the wind and the velocity of the forward sprinting to receive the pass. If the ball is kicked with the correct amount of force and given the right **trajectory**, it drops at the feet of a forward who is running flat out.

The same principles apply to a quarterback who wants to hit a receiver cutting across the field or

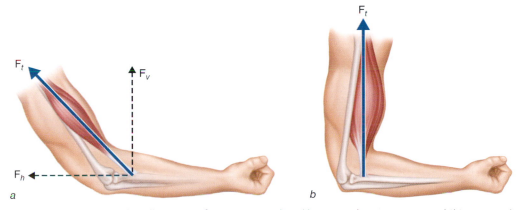

FIGURE 4.3   Force vectors at the elbow joint for (a) a vertical and horizontal component and (b) a vertical component. F_t is the torque force generated by the muscle.

Adapted from S.J. Hall, *Basic Biomechanics*, 4th ed. (Boston: McGraw-Hill, 2003), 166.

a basketball player attempting to hit a teammate who has broken away from the opposition. In all cases, the passers perform a mental vector analysis to make sure that the ball arrives at a particular spot at the same time as their teammate.

When an athlete performs a sport skill, several forces usually act at the same time. Let's look at these forces at work in the shot-put event. Think of elite athletes putting a shot at a release angle of about 42° to the horizontal. To give the shot some height, athletes must apply force in that direction. So the athletes apply some (but not all) of their force in a vertical direction. To get the shot moving horizontally, they apply force in that direction as well. The combination of horizontal and vertical forces gives the shot its 42° trajectory.

Obviously, the shot putters cannot apply all their force in just a vertical or horizontal direction. If an athlete channels all her force in a vertical direction, the shot will naturally go straight up and come straight down. This result is hardly desired in a competition that is won by achieving the greatest horizontal distance. On the other hand, if the thrower directs all her force horizontally, the shot will hit the ground long before it has time to cover the optimal distance. The ideal trajectory angle is partway between horizontal and vertical, or the resultant force.

During flight, the earth's gravitational force pulls the shot directly downward. Gravity fights against only the vertical force vector that the athlete applied to the shot. Gravity is not interested in the horizontal force vector. In addition to the force of gravity, air resistance provides a force that battles the forward motion of the shot. The result of this war of forces determines the distance that the shot travels (see figure 4.4).

Linear Motion for Projectiles

In many events athletes (and objects) are projected through the air to be caught or to land on the ground or in water. Athletes in sports such as ski jumping, trampoline, trapeze, and diving become **projectiles** themselves, whereas objects like golf balls, basketballs, baseballs, and javelins fly through the air and can land or be struck by the sporting implement. These sports all require an athlete to manipulate, control, or assess the flight path that occurs. The key, but subtle, difference when understanding motion is to determine the perspective. When a basketball player shoots the ball toward the hoop, the athlete internally is generating motion, but externally the basketball will experience projectile motion because it will travel through the air. Here are some other examples:

- An archer angles the bow and pulls the strings back just the right amount so that the arrow (as a projectile) flies to the bull's-eye.

- A high jumper aims for height, distance, and rotation so that he (as a projectile) can clear the bar.

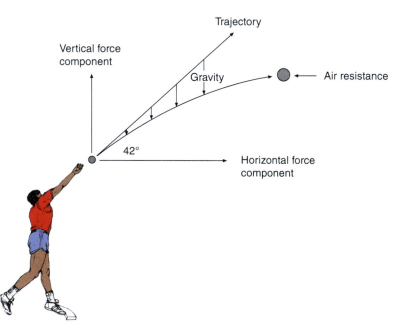

FIGURE 4.4   Forces influencing the trajectory of a shot in the shot-put event.

- A diver (as a projectile) looks for a flight path that gives adequate time in the air to perform all the required twists and somersaults yet still line up for a splashless entry.

- A goalkeeper assesses the velocity and flight of the ball or puck (as a projectile) to make a successful save (with the goalkeeper diving as a projectile).

- A tennis player tries to land the ball (as a projectile) in the correct place in the opposing court but in a location where the opponent is unable to return it successfully.

In events that incorporate flight, several factors are "vectored" in to influence the character of the flight path. An athlete takes off at an angle in jumping events. A baseball is given a trajectory angle during a pitch or when the ball is thrown or hit. Jumpers vary the speed they use during takeoff. And the speed at which a baseball comes off the bat or is released from the pitcher's or fielder's hand varies as well. Finally, you need to consider the height at which the athlete takes off or the height at which a baseball is hit or thrown (see figure 4.5).

Path of the Projectile (Trajectory)

To understand how all these factors are interrelated, let's consider the flight of the baseball (Kageyama

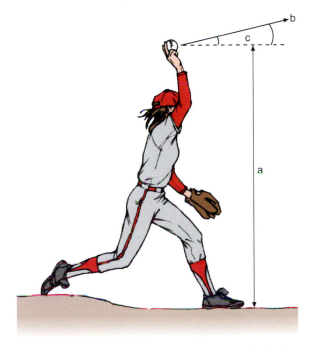

FIGURE 4.5 A baseball pitcher (a) delivers the ball at a particular height and (b) gives the ball a particular velocity and angle of release (c).

Adapted from S.J. Hall, *Basic Biomechanics*, 2nd ed. (Boston: McGraw-Hill, 1995), 313.

et al. 2014). We'll start by eliminating gravity and air resistance, and then we'll get a pitcher to throw a baseball so that it is released at an angle of 35° above the horizontal. To produce the 35° trajectory, the pitcher must apply slightly more force to the ball in a horizontal direction than in a vertical direction. Without the presence of gravity and air resistance, a baseball released at an angle of 35° will fly indefinitely on and upward at the speed the pitcher applied to the ball at the instant of release.

In reality, we all know that gravity (discussed in chapter 7) pulls the projectile (in this case a baseball) toward the earth, and without considering the effects of spin, we also know that air resistance counteracts the forward motion of the ball as it rises and falls in the air.

- Gravity and air resistance change the flight path of the baseball from its trajectory, set by the pitcher at 35°, to the familiar curved flight path in which the ball rises to a certain point and then arcs back toward the earth's surface.

- Gravity battles the rise of the ball upward from the surface of the earth. It finally stops the upward motion and changes it so that the ball falls back toward the earth.

- Remember that gravity pulls perpendicularly toward the earth's surface; it does not counteract the force that the pitcher applied to the ball in a horizontal direction. Gravity battles only the force applied in a vertical direction.

Three key factors influence the linear motion of a projectile:

1. Angle of release
2. Speed of release
3. Height of release

To see how angle of release, speed of release, and height of release combine to influence the distance (or range) that a projectile (for example a baseball or javelin) travels during flight, let's look at each of these factors in turn. The essential factor to create this linear motion in the projectile is determined by the anatomical–mechanical interaction generated by the athlete (Liu et al. 2010).

Angle of Release

When a pitcher hurls a baseball, the shape of the ball's flight path depends on the angle at which the athlete releases the ball. The speed with which the athlete throws the ball then determines the

AT A GLANCE

Producing Linear Motion

- Whenever an athlete performs a sport skill, the athlete primarily produces internal force within the body by contracting the muscles, which results in external movement or linear motion.

- Three key factors influence linear motion of a projectile: angle of release, speed of release, and height of release.

- The key, but subtle, difference when understanding motion is to determine the perspective; is the motion generated internally by the athlete to project something externally, or is the athlete the projectile?

size of the flight path. So the flight path can differ in shape, and each shape can vary in size. If you discount air resistance, the shape of the ball's flight path will be one of three types:

1. If the pitcher hurls the baseball straight up, the ball goes directly upward and gravity pulls it straight down again. The flight path is a straight line. Gravity decelerates the ball on the way up and accelerates the ball on the way down.

2. If the pitcher hurls the ball at an angle between vertical and horizontal, an angle of release that is greater than 45° (i.e., closer to vertical) gives the ball a trajectory in which height dominates over distance.

3. If the pitcher hurls the ball at an angle less than 45° (i.e., closer to horizontal), the flight path is long and low. Distance dominates over height.

Speed of Release

What happens if an object's speed of release is varied? If the pitcher hurls a ball straight up, an increase in the speed of release is obviously going to make the ball go higher. The **apex** of the flight path (the highest point) is raised as the speed of release is increased. The same applies in a vertical jump. The faster the athlete's takeoff speed is, the higher the athlete rises. In each of these cases the athlete's ability to generate speed for the point of release is a consequence of anatomical movements in conjunction with mechanical productivity (Southard 2009). When the pitcher hurls a ball at an angle between vertical and horizontal, any increase in the ball's speed of release increases not only the height of the ball but also how far it travels.

APPLICATION TO SPORT

What Does Movement and Projectile Motion Mean in Sport?

Movement within sport is classified as linear, angular, or a combination of the two, called general. This classification is required so that we can correctly calculate the amount of movement; for example, the linear velocity of a 100 m runner (the velocity as the runner runs straight down the track) is meters per second (or feet per second). By comparison, the angular velocity of the thigh (that is, the velocity of the runner's thigh as it rotates about the hip joint) is degrees per second. The two measures are different, but both are necessary to quantify the sport movement. If an athlete or an implement travels through the air, the movement is governed by projectile motion; the flight path depends on the velocity at takeoff, as well as the angle and the height of takeoff.

Matthias Hangst/Getty Images

Height of Release

The third important factor that influences the flight path of a ball is the height of release relative to the height of the surface on which it lands.

- Golfers most often hit the ball from ground level and frequently have to land it on a fairway or green that is at another level.

- Baseball pitchers and batters release and hit the ball above ground level.

- In track and field, shot putters release the shot well above shoulder level and the shot then lands at ground level. Some shot putters are taller than others, so body type can increase the height of release.

If it were possible for a shot putter to release a shot at ground level and if the ground was perfectly horizontal, then 45° would be the best release angle for the greatest distance. In this situation the athlete puts equal amounts of force in a vertical and horizontal direction. The force applied in a vertical direction is used to battle the downward pull of gravity. When an athlete releases a shot above ground level, however, as all shot putters do, the athlete must lower the trajectory angle and release the shot at slightly less than 45° to get the greatest horizontal distance.

Elite athletes release the shot at an angle that ranges from 35° to 42°, as shown in figure 4.4. This is a large trajectory angle if you compare it with the angle used by ski jumpers when they take off. For example, think of the takeoff of the ski jumper in the same way as the release of the shot. The ski jumper, like the shot, is a projectile. Both fly through the air, and both athlete and shot are projected for maximum distance. Ski jumpers take off from a ramp and land on a surface that slopes downward, similarly to their flight path. So they thrust themselves off at an angle that is close to horizontal.

Long jumpers take off from ground level and want to travel as far as possible, so you might guess that their takeoff angle would be 45°. But this is not the case. These athletes actually take off at an angle between 20° and 22° (see figure 4.6), and the angle of takeoff is even less for triple jumpers. Both types of jumpers would be forced to reduce their speed down the runway to take off at an angle of 45°.

- No long jumper wants to reduce approach speed because doing so would drastically

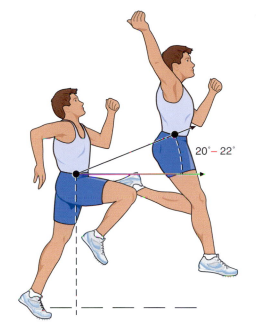

FIGURE 4.6   Takeoff angle in the long jump.

Adapted from E. Kreighbaum and K. Barthels, *Biomechanics: A Qualitative Approach for Studying Human Movement*, 3rd ed. (Upper Saddle River, NJ: Pearson Education, 1990), 394.

reduce the distance traveled in flight and the distance jumped.

- Therefore, long jumpers compromise between velocity at takeoff and takeoff angle.

- Velocity is the more important factor; as a result, the takeoff angle is reduced from 45° to 20° to 22°.

AT A GLANCE

Flying Through the Air

- In many sports, objects (or athletes) are projected or propelled into the air. Their trajectories depend on their velocity, height, and angle of release.

- The forces exerted by gravity and air resistance help determine the resulting flight path.

- With no air resistance, a trajectory angle of 45° produces the greatest distance for objects projected from ground level on a horizontal surface. When the object is projected from above ground level, an angle less than 45° produces the greatest distance.

Velocity, height, and angle of takeoff (or release) are all interrelated. Altering one component inevitably causes changes in the others. Aerodynamic and environmental factors must also be considered. The javelin and discus, for example, are affected significantly by the way that the wind is blowing. Headwinds approaching from a favorable angle and blowing at an optimal velocity (about 24 to 32 km/h, or 15 to 20 mph, for the discus) can dramatically increase the distance that the athlete throws. With a headwind, the thrower not only reduces the angle of release but also lowers the leading edge of the discus relative to air flowing past it. The angle of the discus relative to the airflow is called the **angle of attack**. This concept is discussed in more detail in chapter 7. If the angle of attack is too great, the discus stalls in flight and the distance decreases dramatically. In chapter 10, Moving Through Fluids, you'll read more about aerodynamic factors and the way in which they affect the flight of discuses, javelins, baseballs, and athletes such as ski jumpers.

SUMMARY

- Speed is a scalar measure indicating how fast an object is traveling, measured by dividing the length or distance traveled by the time, but speed does not quantify the direction of travel.
- Velocity is the change in position divided by the time.
- We can measure the velocity of the whole body (for example, the whole body of a runner), or we can measure the velocity of individual components (for example, the velocity of just the runner's leg).
- Acceleration is the rate of change of velocity; an increase or decrease indicates acceleration or deceleration.
- Current-day technologies can be adopted to automate (or semiautomate) the process of measuring linear velocity.
- When the direction and amount of the applied force are known, the combination of these two items is called a force vector.
- Three key factors influence the linear motion of a projectile: angle of release, speed of release, and height of release.
- The key, but subtle, difference when understanding motion is to determine the perspective; is the motion generated internally by the athlete to project something externally, or is the athlete the projectile?

KEY TERMS

acceleration	positive acceleration
angle of attack	projectile
apex	resultant
deceleration	resultant force vector
force	scalar
force vector	speed
ground reaction force	trajectory
linear motion	uniform acceleration
linear velocity	uniform deceleration
negative acceleration	velocity

REFERENCES

Barris, S., and C. Button. 2008. "A Review of Vision-Based Motion Analysis in Sport." *Sports Medicine* 38 (12): 1025.

Kageyama, M., T. Sugiyama, Y. Takai, H. Kanehisa, and A. Maeda (2014). "Kinematic and Kinetic Profiles of Trunk and Lower Limbs During Baseball Pitching in Collegiate Pitchers." *Journal of Sports Science and Medicine* 13 (4): 742-50.

Lee, J.B., K.J. Sutter, C.D. Askew, and B.J. Burkett (2010). "Identifying Symmetry in Running Gait Using a Single Inertial Sensor." *Journal of Science and Medicine in Sport* 13 (5): 559-63.

Liu, H., S. Leigh, and B. Yu (2010). "Sequences of Upper and Lower Extremity Motions in Javelin Throwing." *Journal of Sports Sciences* 28 (13): 1459-67.

Southard, D. (2009). "Throwing Pattern: Changes in Timing of Joint Lag According to Age Between and Within Skill Level." *Research Quarterly for Exercise and Sport* 80 (2): 213-22.

Stamm, A., D.A. James, B.B. Burkett, R. M. Hagem, and D.V. Thiel (2013). *Determining Maximum Push-Off Velocity in Swimming Using Accelerometers*. 6th Asia-Pacific Conference on Sports Technology, APCST 2013, Hong Kong.

www Visit the web resource for review questions and practical activities for the chapter.

5

Linear Kinetics in Sport

When you finish reading this chapter, you should be able to explain

- the fundamental linear kinetic principles—Newton's laws of motion;
- momentum and impulse in sport;
- the factors that influence an object's or athlete's acceleration;
- what is meant by impulse and how impulse relates to sport;
- how an athlete's actions cause reactions that are equal and opposite;
- what is meant by momentum and how objects or athletes gain or lose momentum; and
- how to measure running gait.

You'll learn in this chapter how an athlete's body mass and body weight are related and what is meant by inertia. These fundamental measures are all developed from **Newton's** three laws of motion and create the kinetics in sport (Frost et al. 2010). Imagine an elite sprinter in a set position. The gun goes off, and the sprinter drives out of the blocks. In this situation the sprinter internally applies muscular force that causes his legs to extend, externally creating a force against the blocks. The blocks, of course, are attached to the earth. The force (i.e., the push) that the sprinter applies against the blocks is the action. The reaction (the push back) comes from the earth pushing equally and in the opposite direction, by way of the blocks, against the

sprinter (see figure 5.1). The image in this figure sets the scene for linear kinetics in sport. To promote understanding of this discipline, the chapter explores the principles of linear kinetics, as well as momentum and impulse in sport, and then puts this knowledge into practice by measuring linear kinetics in sport.

Fundamental Principles of Linear Kinetics

Before we begin, we need to brush up on the mechanical principles that are fundamental to understanding applied sport mechanics. The fol-

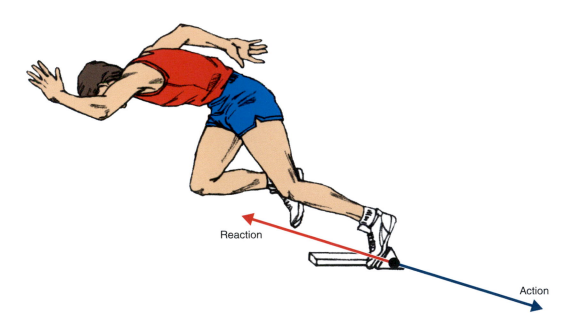

Reaction

Action

FIGURE 5.1   Action and reaction in a sprint start. The athlete applies force against the block. The earth (through the block) applies an equal and opposite force against the athlete.

lowing section provides an overview of Newton's laws of motion and the mechanical terms *mass*, *weight*, and *inertia*.

Newton's First Law of Motion: Inertia

We use the word *inertia* in everyday life to characterize the behavior of people who are slow to commit themselves to action. Therefore, you could say that inertia and laziness are related. In mechanical terms, inertia means more than just laziness because it refers to the "desire" of an object (or athlete) to continue doing whatever it's doing. Inertia means resistance to change. This law of inertia is often reflected as **Newton's first law of motion**.

If an object is motionless, it will "want" to remain motionless. If it's moving slowly it will want to continue moving slowly, and if it's moving fast it will want to continue moving fast. If we are looking at something moving, then the mass of the object will directly relate to the inertia.

- Which is harder to throw, or get moving: a men's shot put (7.3 kg, or 16 lb) or a tennis ball (56 g, or 2 oz)?
- Naturally, the shot put is harder to get moving; the greater the mass that an object has, the more inertia it has too.

We must also consider one more important characteristic of inertia. Once on the move, objects always want to move in a straight line. They will not willingly travel on circular pathways; they need to be pulled or pushed to travel a curved pathway. A ball thrown by an outfielder would travel in a straight line following its release trajectory were it not for air resistance slowing it down and gravity curving its flight path toward the earth's surface.

The more massive an athlete is, the more the athlete's body mass resists change. A giant 136 kg (300 lb) athlete needs to exert great muscular force to get his body mass moving. Once moving in a particular direction, the athlete must again produce an immense amount of muscular force to stop or change direction. Athletes with less body mass have less inertia and therefore need to apply less force to get themselves going. Likewise, they need less force than a more massive athlete to maneuver or stop themselves after they're on the move.

We can identify many examples of inertia at work in everyday life. Oil tankers that cross our oceans have tremendous mass and inertia. They need powerful engines to get them going and huge distances to stop and to turn around. Consider Japanese sumo wrestlers or defensive and offensive linemen in American football. Just like oil tankers, these athletes must apply tremendous force to get their body mass moving and then apply a huge amount of force to change direction or to maneuver the great masses of their opponents.

Newton's Second Law of Motion: Acceleration

Mass simply means substance, or matter. It is typically measured with the units of kilograms (kg) or pounds (lb). People often use *mass* interchangeably with *weight*, but scientifically these terms mean two different things. If an object has substance and occupies space, it has mass. Mass is the quantity of matter that the object takes up.

Weight, on the other hand, is this quantity of matter plus the influence of gravity or, more precisely, gravitational force. So for all our studies on earth, where the acceleration (measured in meters per second squared, m/s^2) due to gravity is pretty constant (at 10 m/s^2), weight is simply mass multiplied by the gravitational force, 10.

- For example, someone with a mass of 100 kg will have a force of weight (measured in newtons, N) of 1,000 N.
- So for coaches, athletes, and sport scientists, mass is the most common term we use, and weight is the force that this mass generates.

We frequently talk of National Football League (NFL) linemen as being massive or having tremendous body mass, indicating that the athletes are enormous and have plenty of muscle, bones, fat, tissue, fluids, and other substances that make up their bodies (Robbins et al. 2014). Athletes who want to perform well in their chosen events carefully monitor their body mass. They know that too much or too little mass can seriously affect their performance.

The acceleration of any sprinter's body mass is proportional to how much muscular force the athlete applies and the period in which the force is applied. It is also inversely proportional to the athlete's mass.

- Therefore, if two sprinters apply the same muscular force to their bodies for the same amount of time, the less massive of the two athletes accelerates more.
- Likewise, if two sprinters have the same mass and apply force for the same amount of time, the athlete who applies more force within that period will accelerate more.

This scenario is a good example of Isaac **Newton's second law of motion**, the law of acceleration, and as a simple formula it is written this way:

$$F = m \cdot a \text{ (force = mass} \times \text{acceleration)} - \text{the same way}$$
we calculated weight from a given mass.

For all of us, checking our body mass is a means of assessing our general health and fitness. When we get on a scale, the dial gives us a reading that we associate with the amount of body mass that we carry around. A common assumption is that an athlete's body mass compresses the springs in the scale and that the readout on the dial represents the amount by which the springs are squeezed together. This interpretation is true, but what actually happens is a little more complex, as discussed next.

In mechanical terms, an athlete's weight represents the earth's gravity pulling on the athlete's body. The readout on the scale represents how much pull or attraction exists between the two. The earth pulls the athlete downward. Therefore, an athlete with more body mass compresses the springs more than an athlete who has less body mass. As a result the needle on the scale moves farther around the dial.

Newton's Third Law of Motion: Action–Reaction

Going back to the sprinter at the start of the chapter, the forces produced by the sprinter's muscles overcomes the inertia of his mass, and he begins to accelerate. If no opposing forces were acting against the sprinter, then his legs extending against the blocks would cause them to continue moving indefinitely in the direction that their muscular force propelled them. Gravity, friction, and air resistance apply the brakes to this endless motion; these additional components are discussed in chapters 8, 9, and 10.

By extending his legs powerfully at the start, a sprinter contracts his muscles to push against his body mass and simultaneously against the mass of the earth, through the blocks. The athlete moves in one direction, and the earth (plus the blocks) moves a negligible amount in the opposing direction. The earth and the athlete move in opposite directions relative to their mass and their inertia.

Put another way, you can picture the relationship between the sprinter and the earth by compressing a spring between a 7.3 kg (16 lb) shot (resting on the ground or another hard surface) and a table tennis ball. The shot is the earth, the spring is the athlete's muscles, and the table tennis ball is the athlete. If you let go of the shot and the table tennis ball at the same time, the spring expands. The table tennis ball accelerates in one direction, and the shot rolls a short distance in the opposing direction, or maybe barely rolls at all. Now visualize increasing the size of the table tennis ball so that it's equal to that of the sprinter and increasing the size of the shot so that it's equal to that of the earth. For that reason, the sprinter moves in one direction relative to his mass (and his inertia) and the earth moves an immeasurable amount in the opposing direction relative to the earth's mass and its inertia. This example represents **Newton's third law, the action–reaction law**. It tells us why the sprinter does the moving!

AT A GLANCE

Linear Kinetic Fundamentals

- Newton's first law of motion is the law of inertia; that is, an object at rest stays at rest, and an object in motion stays in motion.

- Newton's second law is the law of acceleration; that is, the acceleration of an object is directly proportional to the magnitude (and in the same direction) of the net force applied, and can be represented by $F = m \cdot a$.

- Newton's third law is the law of motion; that is, there is an equal and opposite action–reaction to describe motion.

Momentum and Impulse in Sport

Momentum plays a particularly important role in sport and in situations where athletes or objects collide. An easy way to think of momentum is to see it as a weapon that an athlete can use to cause an effect on an object or opponent. When two players on the football field collide, the player with the greatest momentum will keep moving forward. Players can generate momentum by their own combination of mass and velocity, so the heavier player will not always have the greatest momentum. A lighter player, running faster, can generate greater momentum and come out on top of a heavier opponent.

Impulse is the combination of time and velocity of this force. By cleverly manipulating these components in sport, an athlete can enhance a performance or make it safer (or both). If we look at catching a baseball, the velocity of the ball means that it will generate a large momentum. Players attempting to catch the baseball can manipulate the time over which the force is applied to their hand (impulse). They can increase the time by skillfully allowing the hand to move slowly back at the time of **impact**. This longer time reduces the impulse force at impact and can reduce injuries to the hand.

Fielders in the game of cricket don't wear gloves like those used in baseball. Yet a cricket ball is similar in size to a baseball, but harder. To reduce the sting of catching a hard-driven ball, fielders in cricket reach out to catch the ball but at the instant of contact quickly draw the hand backward. This action increases the time of contact and reduces the impulse force felt at the hand.

Momentum in Sport

Like *inertia*, *momentum* is a commonly (although sometimes incorrectly) used term. Sport commentators frequently talk of a team's gaining momentum, and by this they mean that one team is starting to dominate the other. Politicians also talk of their campaigns' gathering momentum. In mechanical terms, momentum has a far different meaning. An athlete who is moving is an example of mass on the move. Because the athlete's body mass is moving, we say that the athlete has a certain amount of momentum.

Momentum describes the quantity of motion that occurs. How much momentum an athlete possesses depends on the mass of the athlete and the speed at which the athlete is traveling at the time. An increase in the athlete's mass, velocity, or both increases the athlete's momentum. The simple formula for momentum is written:

$$M = m \cdot v \text{ (momentum = mass} \times \text{velocity)}$$

- Therefore, momentum is a direct function of mass and velocity. If movement is not occurring, neither is velocity, and naturally there is no momentum.

- Likewise, a massive athlete and a smaller athlete could develop the same amount of momentum, depending on the velocity at which they are traveling.

- To make up for a large difference in mass, athletes with little body mass must sprint at a much higher velocity to match the momentum of a more massive athlete.

As an example, if a 136 kg (300 lb) lineman ambled through 100 m in 20 s (which is really slow!), an athlete at half that mass, say a 68 kg

---

APPLICATION TO SPORT

Mass, Weight, and Inertia in Sport

The mass of athletes reflects how heavy they are. In some sports, such as NFL football, having more mass makes athletes more effective. Having greater mass means that they have greater resistance to a change in their state of motion (either stationary or dynamic), namely their **inertia**. Therefore, a heavier player has greater resistance to a change in his state of motion, which is useful in blocking and in stopping the opposing players from getting through. This mass and inertia relationship is found in any activity in which an athlete or an object (like a bowling ball) is required to change the state of motion; it therefore plays an important role in, first, understanding what is happening in sport and, second, knowing how to enhance this activity. As mentioned earlier, weight is a function of mass and is simply the gravitational force of mass. So on earth, weight is the mass of the object (human or implement) multiplied by the gravitational force, which is approximately 10 m/s². We use the term *weight* to calculate the force generated, and we use the term *mass* to calculate how heavy something is.

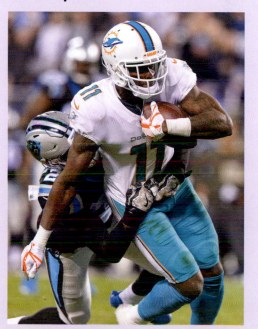

Grant Halverson/Getty Images

(150 lb) running back, would have to flash through the same distance in 10 s to produce the same momentum.

We have seen that if you increase either the velocity or the mass of an object or an athlete, momentum increases as well. If there is no velocity, there is no momentum. The best way for an athlete to increase mass is to pack on quality muscle mass rather than fat. The extra muscle mass provides the power to help the athlete move faster and maneuver more efficiently.

Remember that not all sport situations require maximum momentum (Kageyama et al. 2014). Many skills require careful control of momentum. For example, a punter often has to make the ball go out of bounds as close as possible to the end zone. The momentum given to the ball has to be exact so that it travels a precise distance. This means that the momentum of the punter's kicking leg must be the right amount, not necessarily the maximum. Similarly, an outside shooter who fires at the basket from three-point range aims to put the basketball through the hoop, not have it ricochet haphazardly out to midcourt. The trajectory, spin, and momentum given to the ball must be exact when it is released.

Impulse in Sport

For an athlete to make something move, say a soccer ball, an internal muscular force of the leg is required to accelerate the ankle so that this moving object has momentum before the collision with the soccer ball. If we look at this collision in more detail, the force that the athlete applies is exerted over a certain period. When athletes apply force to an object over a certain time, we say that the athletes have applied an **impulse** to the object. Of course, athletes can also apply an impulse to their own bodies or to another athlete.

How force and time are combined depends on the physical capabilities of the athlete. An athlete who is strong and flexible can apply more force over a greater range (i.e., period) than an athlete who is weaker and less flexible. Equally importantly, the combination of force and time depends on the needs of the skill. Some skills, such as those used by a boxer in delivering certain punches, require tremendous force to be applied over a short distance and a short time. Other skills require less force to be applied over a longer time. The variations in the use of force and time are limitless. Let's look at some examples to see how the requirements of sport skills vary in the demand for force and the time of its application.

Impulse in a Javelin Throw

A javelin throw is an example of the application of incredible force to an implement over a long period. Power and flexibility are required in this event. After an approach run, an expert thrower accelerates the javelin by pulling it from way behind his body and releasing it far out in front (Liu et al. 2010). Long arms are beneficial, but more important is the backward body lean when the athlete enters the throwing position. In this way, the athlete can apply the force to the javelin over a longer period.

To a spectator it may not seem that the athlete is accelerating the javelin for long because the whip-like action of a good javelin throw seems to happen so quickly. But when an elite thrower is compared with a novice, the distinction between the two athletes is easy to see, even for someone not familiar with the event. To start with, the elite thrower is much more powerful and applies more force to the javelin at high speed. Second, greater flexibility and careful technical training allow the elite athlete to accelerate the javelin over a longer time. Therefore, the impulse applied to the javelin by an expert thrower is far greater than that applied by a novice, and as a result the javelin moves at tremendous velocity when it is released (see figure 5.2).

Impulse at Takeoff in the High Jump

High jump is similar to javelin in that both events require the athlete to generate considerable velocity. The javelin is a projectile hurled by the thrower, and a high jumper becomes a projectile propelled upward into the air by muscular force. Because the high jumper wants to go as high as possible, you might think that applying as much force as possible at takeoff over the longest available period would be beneficial. So why don't we see athletes lowering down onto a fully flexed leg and then thrusting upward until the jumping leg is fully extended? Surely this would maximize the total force applied by the leg muscles.

Unfortunately, this is not the case. Starting from a fully flexed position, the athlete cannot develop maximum force because the leg muscles in the jumping leg are in an anatomically poor position to drive the athlete upward. What you'll find instead is that all great high jumpers start their upward thrust from a jumping leg that is flexed no more than the equivalent of a quarter squat.

- If high jumpers cannot use a jumping leg that is fully flexed at the knee, is there any

FIGURE 5.2   Impulse in a javelin throw. Elite athletes are able to apply force over a long period by (a) leaning back and pulling the javelin from behind the body and (b) releasing it in front of the body.

other way they can extend the time over which they apply force? Yes, there is.

- Like elite javelin throwers, all great jumpers lean backward as they plant the jumping foot before takeoff.
- Straightening up from a backward lean allows the athlete to spend more time applying force to the ground, which in reaction thrusts the athlete upward (see figure 5.3).

We know from sport science assessments that the same technique is used by volleyball players when they jump to spike and block, by soccer players when they jump to head the ball, and by basketball players when they leap to block or perform a layup. The use of a backward lean before takeoff increases the time that the force is applied (impulse); therefore, for the same amount of force production, a longer impulse results in a higher flight. Watch slow-motion videos of great ballet performers such as Mikhail Baryshnikov and Rudolf Nureyev when they jump upward. The correct mechanics of getting up in the air in ballet are no different from those required of a high jumper, basketball player, or volleyball player.

FIGURE 5.3   Impulse in a flop high-jump takeoff. (a) Elite high jumpers lean back before takeoff, which (b) allows them to spend more time applying force to the earth. The earth, in reaction, thrusts the athlete upwards.

When generating impulse for the high jump, the rules of high jumping don't reward an athlete who can jump the highest in any style, but rather the athlete who can jump the highest off one foot (and then cross a bar). If a two-foot takeoff were allowed, the world record of just over 2.4 m (8 ft) would surely be beaten by a gymnast using a high-speed run-up, a round-off, and several back handsprings to gain speed, followed by a triple back somersault to cross the bar. Today's elite gymnasts reach heights of 2.7 to 3.0 m (9 to 10 ft) on the second of the three somersaults. Forty years ago, using a single back somersault, gymnasts cleared a bar set at 2.28 m (7 ft 6 in.). As you can see, different rules are required in high jumping and gymnastics to ensure that the skills of the sport are maintained.

Impulse and Cadence in Sprinting, Speed Skating, and Rowing

The sports of sprinting, speed skating, and rowing are similar in that athletes perform the same actions in a cyclic, repetitive fashion throughout the race. These skills are not like the high jump (in which the athlete can rest after a jump), or the volleyball spike (which can be followed immediately by other skills like blocking, digging, or setting).

- Elite track athletes repeat their running action for the duration of their race.
- Rowers repetitively pull on the oars.
- Speed skaters thrust and glide at high speed around the rink.

Throughout their races these athletes vary the amount of force and the time over which they apply force with their muscles. Because they are accelerating and overcoming their own inertia, rowing eights use a higher cadence with more strokes per minute at the start than they do farther down the course. Each pull on the oar is quick and powerful, but over a short range or distance.

Likewise, sprinters and speed skaters use short, quick strides as they accelerate from the start. Once moving at high velocity, they reduce their stride rate, but each stride is longer. Why? The answer is best explained scientifically. The most efficient way of overcoming inertia is to apply great force quickly and repeatedly over a short distance, or within a short range of motion. This method has been found to be the best way for a rowing eight, a sprinter, or a handcyclist (Abel et al. 2015) to accelerate and get up to top velocity as quickly as possible. Unfortunately, a high stroke or stride rate burns up a lot of energy and, although this method is efficient when athletes are accelerating, is inefficient after they are moving at high velocity. Once up to speed, sprinters and speed skaters

---

APPLICATION TO SPORT

Absorbing the Impact

Air bags in cars act like a pole-vaulter's landing pad: Both are designed to absorb impact. A pole-vaulter's landing pads are permanently filled with absorbent material that cushions the athlete's landing. Air bags in cars have to fill at high speed to absorb the impact of a driver or passenger who is thrown forward in a collision. Using the same principle as the air bags in cars, air bubbles cushion springboard and tower divers in training. Pressurized air is released from the bottom of the diving tank. The air expands as it rises and provides an elevated area of frothy water at the surface. A failed dive causes less punishment because the athlete lands on a bed of watery air bubbles. Divers can fail in a dive from the 10 m tower knowing that at over 48 km/h (30 mph), most of the sting of hitting the water has been eliminated.

Simon Bruty/Allsport/Getty Images

reduce their stride rate and extend more fully with each leg thrust.

Because of the mechanical advantage of the oar, rowers pull over a larger distance with each stroke using a greater range of motion. Therefore, even though stride rate and stroke rate may be reduced, great force is applied over an increased range of motion at a lower cadence. This reduction in cadence helps the athletes maintain velocity without running out of energy.

Using Impulse to Slow Down and Stop

Let's now look at impulse in a different light. In this section you'll see how impulse is used to slow down and stop an object. Consider a basketball player who leaps for a slam dunk and afterward lands stiff legged on the floor. The player's mass, dropping from a height of say 3.0 m (10 ft), slams into the floor and comes to a halt in an instant. The reaction force of the floor (i.e., the earth) hits back at the athlete with the same force at which the athlete hits the floor.

- Unfortunately, the time during which the athlete's body must absorb this force is extremely short, so the shock and stress on the athlete's body are phenomenal.
- What do most athletes do naturally to counteract this? They flex at the ankles, knees, and hips.

Coaches often tell their athlete, "Bend your legs as you land!" (Gehring et al. 2009). In a mechanical sense, this coaching advice tells the athlete to extend the time during which the body receives and absorbs the force applied by the ground. At any instant over this longer period, the force applied to the athlete's body will be less.

Besides lengthening the period over which force is applied to their bodies, athletes learn to enlarge the area of impact (i.e., the place where forces are applied) as much as possible. A runner's slide into home plate is a good example. The slide not only gets the runner's legs below the reach of the opponent's tag but also extends the time during which friction with the ground brings the athlete to a halt. In addition, the sliding action puts a large area of the runner's body in contact with the ground.

Visualize the pain and discomfort of a base runner who dives at the bag headfirst and stops in an instant on the point of his nose! Although this comical scenario seldom, if ever, occurs, it illustrates a situation in which all the forces produced by the athlete are reduced to zero in an instant, and in a small and sensitive area.

Athletes in many sports are taught specific techniques to extend the area and time over which forces act on their bodies. By doing this they avoid injury and reduce to comfortable levels the pressure exerted on them. Ski jumpers perform a telemark landing after flying through the air, as shown in figure 5.4. Flexing their legs with one leg forward and one leg back lengthens the time that the impact force of landing is applied to their bodies.

FIGURE 5.4   Reducing the impact of landing in ski jumping.

Hockey players try to ride out the force of a check from an opponent, and athletes in judo use breakfall techniques that enlarge the area and the period over which the force of impact with the mat (i.e., the earth) is applied to their bodies. Perhaps one of the most famous examples of this mechanical principle was Muhammad Ali's legendary technique of rolling with the opponent's punch. As his adversary threw a punch, Ali rolled his head and body backward. In this way the force of the opponent's punch was extended over a time and therefore had less impact and effect. Imagine if Ali had stepped forward into a punch thrown by his opponent. The period of contact would have been reduced to an instant, and the effect of the punch would have been much greater.

Athletes do not always have to rely on special techniques to avoid injury when they are involved in impact situations. They get help from equipment that is designed to extend the time and enlarge the area over which external forces are applied to their bodies. Helmets, padding, gloves, crash pads, foam-rubber-filled landing pits, and even parachutes do this job. The total force applied to the athlete's body does not change, but when the time and the area of application are extended, the force applied at any one instant and in any one place on the athlete's body is significantly reduced. Air bags in cars use the same principle.

How to Measure Linear Kinetics: Running Gait

Running is a fundamental component in almost every land-based sporting activity, so in this section we will look at how to measure the intralinear kinetic components for the common human activity of running. If you look back at figure 4.1, graphing the components of Johnson's sprint, you'll see that **stride length** and **stride rate** (sometimes called **stride frequency** or **cadence**) are mentioned. The stride time is the time taken for each stride. Here we will examine how to measure these characteristics, or **temporal gait characteristics**.

- The formula for running velocity is stride length times stride rate (running velocity = stride length × stride rate).
- Stride length is defined as the distance between successive points of initial contact (usually called foot contact or heel strike) of the same foot, and right and left stride lengths are normally equal. **Step length** is the distance between the point of initial contact of one foot and the initial point of contact of the opposite foot. In normal gait, right and left step lengths are similar.
- Stride rate is the frequency or cadence of the leg swings.

When walking or running, the athlete alternates between the

- **stance phase** (when the foot is making contact with the ground) and the
- **swing phase** (when the foot is swinging through the air).

The stance and swing phases are the typical parameters that define gait. In walking, the stance phase normally accounts for about 60% of the gait cycle and the swing phase 40% (for running, these percentages vary according to the velocity). The identification of these features is important to understanding and quantifying how an athlete is moving. Some typical mean temporal values for walking gait are shown in table 5.1. These values vary for gender, for children through to adults, and depending on factors such as leg length.

How an object is moving can be measured in a number of ways. In particular, advances in technology enable movement to be quantified and presented uniquely. This understanding can help us determine whether an athlete is moving effectively or whether an intervention (such as using a different type of golf club) can improve performance. One of the most accessible methods is video analysis, which allows a permanent record of the event to be made. From this record, a number of sport mechanics calculations can be obtained. This record is also a useful tool to show athletes what they are doing during the activity.

Often, when athletes can see their own performance, they can then (finally) comprehend the feedback that the coach has been providing. The video also allows development of a progressive time line comparison. For example, the record of a performance last season can be compared with that of the current technique, and this progression can be a powerful reinforcement to the athlete that all the training is effective. To use video for qualitative movement analysis, a few simple steps are followed, as discussed in the next sections (we will explain using the measurement of running gait as an example).

TABLE 5.1 Walking Patterns of Men and Women

Variable	Men	Women
Step length (cm)	79	66
Stride length (cm)	158	132
Cadence (steps/min)	117 (60-132)	117 (60-132)
Velocity (m/s)	1.54	1.31

Adapted from M.P. Murray, A.B. Drought, and R.C. Kory, 1964, "Walking Patterns of Normal Men," *Journal of Bone and Joint Surgery* 46(2):335-360, and M.P. Murray, R.C. Kory, and S.B. Sepic, 1970, "Walking Patterns of Normal Women," *Archives of Physical Medicine and Rehabilitation* 51: 637-650.

AT A GLANCE

Linear Kinetics: Running Gait

- Running is a fundamental component in almost every land-based sporting activity.
- Running velocity is stride length times stride rate (running velocity = stride length × stride rate).
- When walking or running, the athlete alternates between the stance phase (when the foot is making contact with the ground) and the swing phase (when the foot is swinging through the air).

Setting Up the Video Camera

The following steps will enable a repeatable measure using the video camera:

1. Mount the camera in a fixed position on a rigid tripod.

2. Place the camera so that the axis of the lens is perpendicular to the direction in which athlete will be walking or running (this direction is often called the plane of motion).

3. Place the camera as far away from the motion as is practical to avoid parallax errors (if the camera is too close to the object, the outside edge of the image captured will not be parallel; increasing this distance will reduce the parallax error).

4. Zoom the lens so that the action fills the field of view without cutting out critical body parts.

5. Level the camera using a bubble level on the tripod or camera.

6. Ensure that the center of the field of view (height and width) is at the center of the desired motion.

7. Focus the lens.

8. If the camera has an adjustable shutter, reduce the exposure time to as short a period as possible without compromising the picture brightness (note that extra lighting may be required and that a short exposure time reduces the blurring of a fast motion when a video frame is paused).

9. Display the date and time on the camera, if possible. This information helps to reduce digitizing time and helps keep track of data.

10. Record a few frames of a meter stick held in the plane of motion. This step is required to allow calibration of the measurements to be made after digitization.

Digitizing and Making Measurements From Video Data

After the activity has been captured on video, you can present the data in several ways. First, playing back the video to the athlete can provide a global and normal-speed feedback. But to identify particular features of the performance, the slow-motion or frame-by-frame advance (or both) is an effective way to break down what is happening. These functions also allow measurements, such as stance and flight time, to be made. The following steps provide the process to measure the stance or swing time during running (or walking).

1. To play back the video image, use the LCD screen on the camera or, if a larger view is required (a better option to allow the athlete to view the image), connect the camera to a TV or computer.

2. To determine the stride rate (cadence or frequency), play the video until one foot is about to make contact with the ground. By using the frame forward or backward option on the camera, you can identify the exact frame in which the foot makes contact with ground. Make note of the video time clock for this frame, which will be the baseline reference point.

3. Using the frame forward button, advance the video until the same foot leaves the ground. Make note of the video time clock; the difference in time between this frame and the baseline reference point is the stance time for that leg.

4. Using the frame forward button, advance the video until the same foot makes contact with the ground again. Make note of the video time clock; the difference in time between this frame and the baseline reference point is the stride time for that leg.

5. To determine the step time, you can repeat the process just described, but for step rather than stride.

6. To determine stride length, we can either walk or run for a set number of strides (for example, 10) and, using a tape measure, record the distance covered from the start

to the finish for the 10 strides. The stride length is then the distance covered divided by 10. To measure stride length using the video, we first need to calibrate the meter stick, as mentioned previously.

To digitize a known length (a meter stick, for example), follow the process outlined here. When complete, this process allows a number of other measurements to be made from the stationary video image. You can digitize in a couple of ways. You can make measurements on either a TV or a computer screen, or, if the video can be imported into a third-party software system in the computer, you can use a mouse to make the measurements. Either way, the fundamentals of the process are to calibrate the known length (meter stick) into the measurement made on the screen.

1. If you are using a TV or computer screen to play back the image, cue the video to the captured image of the object with a known length (meter stick).

2. To protect the TV or computer screen, place a clear sheet of plastic on the screen and tape it down so that it will not move.

3. Using a ruler, measure the length of the image of the known object. For example, on the TV screen, the image of the 1 m ruler could be 0.325 m. This ratio, of TV image distance divided by known distance (0.325 / 1 = 0.325), is then applied to any measurement we now make on the TV screen. For example, using the ruler on the screen, if we measure the distance from foot contact to foot contact as 0.260 m, the actual distance is this measured value divided by our ratio. This would be 0.260 / 0.325, which equals 0.8 m. By calibrating the video image, we can calculate the stride length measurement in this example as 0.8 m.

APPLICATION TO SPORT

The Cost of Slowing Down in a 100 m race—a World Record?

The blistering-fast world-record time of 9.69 s for the 100 m sprint by Usain Bolt at the 2008 Beijing Olympics was an amazing performance, taking 0.03 s off the mark he had set in May that year. Watching the race, it was evident that Bolt started slowing down at around the 80 m mark, and the question is raised—what could the world record have been if he had not slowed down? By calculating Bolt's early speed, acceleration, and position before effectively switching off and slowing down, we can model and predict the final velocity and subsequently the time to run 100 m. Several people have calculated and predicted the possibilities. Norwegian physicist Hans Eriksen estimated that the Jamaican athlete could have clocked between 9.55 and 9.61 s in China if he hadn't slowed down to celebrate his gold medal effort with about 20 m to go. Usain Bolt set the current world record at 9.58 s one year later in 2009. The women's 100 m world record of 10.49 s was set by Florence Griffith-Joyner way back in 1988.

Cameron Spencer/Getty Images

SUMMARY

- Newton's first law of motion is the law of inertia; that is, an object at rest stays at rest, and an object in motion stays in motion.
- Newton's second law is the law of acceleration; that is, the acceleration of an object is directly proportional to the magnitude (and in the same direction) of the net force applied.
- Newton's third law is the law of motion; that is, there is an equal and opposition action–reaction to describe motion.
- The acceleration of athletes or objects is proportional to the force applied against them and the time during which this force acts. An increase in force, time, or both will increase acceleration.
- Force multiplied by the time over which it acts is called impulse.
- All actions cause reactions. Action and reaction are colinear forces, being equal as well as acting in opposite directions. The law of action and reaction is Isaac Newton's third law.
- Momentum describes quantity of motion. An increase in mass, velocity, or both increases momentum.
- Force that is absorbed over a long period and over a large area helps prevent injury when athletes come to a stop or when they stop moving objects.
- The temporal gait characteristics of step length, stride length, stride rate (cadence), and swing and stance time are useful mechanisms for quantifying walking or running gait.

KEY TERMS

cadence

impact

impulse

inertia

momentum

newton

Newton's first law (the law of inertia)

Newton's second law (the law of acceleration)

Newton's third law (the law of action and reaction)

stance phase

step length

stride frequency

stride length

stride rate

swing phase

temporal gait characteristics

REFERENCES

Abel, T., B. Burkett, B. Thees, S. Schneider, C.D. Askew, and H.K. Strüder. 2015. "Effect of Three Different Grip Angles on Physiological Parameters During Laboratory Handcycling Test in Able-Bodied Participants." *Frontiers in Physiology* 6 (November): 331.

Frost, D.M., J. Cronin, and R.U. Newton. 2010. "A Biomechanical Evaluation of Resistance Fundamental Concepts for Training and Sports Performance." *Sports Medicine* 40 (4): 303-26.

Gehring, D., M. Melnyk, and A. Gollhofer. 2009. "Gender and Fatigue Have Influence on Knee Joint Control Strategies During Landing." *Clinical Biomechanics* 24 (1): 82-87.

Kageyama, M., T. Sugiyama, Y. Takai, H. Kanehisa, and A. Maeda. 2014. "Kinematic and Kinetic Profiles of Trunk and Lower Limbs During Baseball Pitching in Collegiate Pitchers." *Journal of Sports Science and Medicine* 13 (4): 742-50.

Liu, H., S. Leigh, and B. Yu. 2010. "Sequences of Upper and Lower Extremity Motions in Javelin Throwing." *Journal of Sports Sciences* 28 (13): 1459-67.

Robbins, C.A., D.H. Daneshvar, J.D. Picano, B.E. Gavett, C.M. Baugh, D.O. Riley, C.J. Nowinski, A.C. McKee, R.C. Cantu, and R.A. Stern. 2014. "Self-Reported Concussion History: Impact of Providing a Definition of Concussion." *Open Access Journal of Sports Medicine* 5:99-103.

www Visit the web resource for review questions and practical activities for the chapter.

6

Angular Motion in Sport

When you finish reading this chapter, you should be able to explain

- the factors necessary to initiate and vary angular motion;
- mechanical principles for rotation;
- how athletes apply torque and how torque varies angular motion;
- why rotating objects initially resist rotation and then want to continue rotating after they've been set in motion;
- how the resistance of an object to rotation can vary; and
- how to measure angular rotation.

Angular motion in sport is naturally about rotation. This chapter is the second longest in this book; only the complex topic of moving through fluids is larger. The length of this chapter is an indication not only of the importance of rotation but also of its presence in all sport skills. The first half of this chapter explains the basic principles that relate to rotation, or **angular motion**. When looking at the anatomy of the human, we see that all our limbs rotate about the center of axis for that joint, so rotation is important for understanding human movement. As with the previous chapters this knowledge is put into practice with applications in a number of sports, in particular measuring the angular rotation of the lower limb during running. In applied sport mechanics, angular motion describes an athlete rotating, circling, revolving, spinning, somersaulting, twisting, pirouetting, turning, and swinging. All these terms refer to angular motion in sport.

Fundamental Principles of Angular Motion

If you watch closely, you'll see that rotation occurs in all sports, including those that require an athlete to stay as still as possible. In Olympic pistol and rifle shooting, for example, athletes try to eliminate all unnecessary movements. They train to slow down their heart rate and, amazingly, to pull the trigger between heartbeats so that the thump of the heart doesn't jolt the barrel of the gun. Even in this slow and patience-demanding sport, the bones of the index finger move in a rotary fashion when the athlete squeezes the trigger. An archer does the same thing when she flexes her arm to pull a bow. As the archer draws back the bowstring, the forearm and upper arm flex and rotate toward each other. Both shooter and archer hardly move at all, but rotation occurs nevertheless.

Far different from archery and shooting are the sports of gymnastics, diving, ski aerials, and figure skating. Dramatic rotary skills performed in flight characterize these sports. Because they compete in so many different events, gymnasts perform the greatest variety of rotational skills. In contact with stable apparatus (e.g., the floor, beam, vault, bars, and pommel horse) and highly unstable apparatus (i.e., the rings), gymnasts perform somersaulting and twisting skills around axes formed by their feet, hands, hips, shoulders, and even their knees.

The prize for the greatest number of somersaults and twists performed at any one time belongs not to gymnasts but to ski aerialists. These daredevil athletes use a "kicker" (i.e., a ramp) that throws them high in the air. With their flight time extended by a downward-sloping "outrun," they combine as many as three somersaults with five twists.

An athlete or an object rotates, spins, swings, or twists through an angle of a certain number of degrees using angular motion.

- If a gymnast makes one full revolution around a bar, she has rotated through an angle of 360°. In golf, from backswing to follow-through, a golfer can swing a club from 15° in a putt to a complete 360° when driving.
- Basketball players who perform "360s" in slam-dunk competitions spin one full revolution around their long axis before slamming the ball through the hoop.
- Snowboarders in the half-pipe can perform "720s" as they rotate their whole body and snowboard on the snow.

Whether rotation occurs through several revolutions or through an arc of a few degrees, the same mechanical principles apply. Understanding these principles is helpful because they will help you teach technique based on sound sport mechanics.

The fundamental components for anything to rotate are an axis of rotation (fulcrum or pivot point) and a lever attached to this reference point.

- The simplest rotation device is a bicycle pedal. The axis of rotation is the center of the pedal, and the lever is the pedal crank.
- This same principle of axis of rotation and a lever applies to all rotations.
- The effect of this rotation is often measured from the object's (human or implement) center of gravity. This component plays a key role in sport mechanics, as demonstrated the next chapter, on the topic of angular kinetics in sport. Briefly, if we collectively add up the weight from each of the parts of the object (in a human this would be the legs, torso, arms, head, and so on), we can find a **balance point**, the point at which the force of weight is equal on either side. This balance, or center point, is the object's center of gravity.

When we look at human rotations, just as with all other movements, we can view them as occurring in three orthogonal (mutually independent) planes, as shown in figure 6.1:

- The **longitudinal axis**, where rotation occurs around the length of the body, is the major axis in the human body. For example, imagine twisting a long barbecue skewer to cause rotation, similar to a gymnast's twist.
- The **transverse axis** occurs from side to side or from hip to hip. For example, imagine a gymnast's somersault.
- The **frontal axis** occurs from front to back. For example, imagine a gymnast's cartwheel and side somersault.

When any object rotates, a number of factors come into play, like the Magnus effect of spin on a ball or object, which is covered in chapter 10, Moving Through Fluids. To determine what happens, we first need to quantify how much rotation is occurring. The amount of rotation is influenced by velocity, force, inertia, and momentum. Each of these is discussed in the following sections.

Angular Velocity

Angular velocity is a term used to describe the rate of spin of an athlete or an object. Just as linear velocity describes the rate of movement in a linear direction, angular velocity describes movement in an angular direction. The units of measurement for angular velocity are degrees per second or, for faster movement, radians per second.

- One radian is 57.3° (which is easy to remember, because this value is 180° divided by pi, or π; 180 / 3.14 = 57.3).

Angular velocity can then describe the rate of swing of a bat or club. But the angular velocity and the speed of an object or an athlete as the object or athlete rotates are different. To understand this difference, consider a gymnast performing giants (i.e., 360° rotations) around the high bar and rotating around the bar with a perfectly rigid body. In reality, gymnasts must flex at the hips and at the shoulders at specific phases in a giant around the high bar; but for this example, we'll assume that the gymnast remains rigid throughout this skill (Hiley et al. 2015). We will discuss this concept in more detail in the next section on applied angular motion.

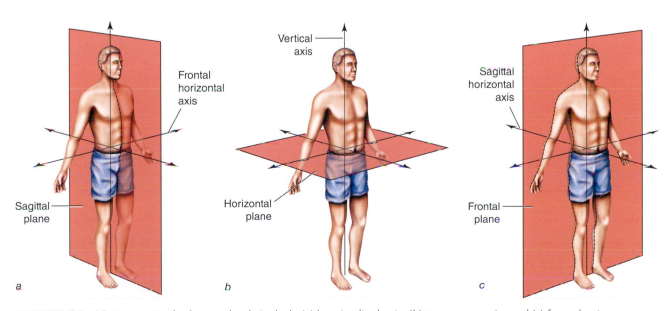

FIGURE 6.1   Major axes in the human body include (*a*) longitudinal axis, (*b*) transverse axis, and (*c*) frontal axis.

If you time the gymnast as he rotates around the bar, your stopwatch might show that he makes one complete revolution in a clockwise direction every second. You now know the angle, or the number of degrees (i.e., 360° for each revolution), that the gymnast performs in a particular time and in a particular direction. If you know the number of revolutions, the period, and the direction in which the gymnast is rotating, then you know his angular velocity, or rate of spin (angular refers to angle, degrees, or revolutions; velocity means speed with direction). All parts of the gymnast's body make one complete circuit around the bar in a clockwise direction in 1 s.

Let's now turn our attention to the gymnast's body as it rotates. If you watch his hips as they follow their circular pathway and then watch his feet, you'll come to the following conclusions:

- The gymnast's hips travel around a much smaller circle than his feet. His feet, which are farther from the bar, travel around a bigger circle.

- The gymnast's hips and feet complete their different-size circles in the same time. So the gymnast's hips and feet have the same angular velocity.

- If the gymnast's feet are twice the distance from the bar that his hips are, they travel around a circle that is twice as big and as a result must be moving twice as fast. Their speed is twice that of the hips.

This information tells us that although all parts of the gymnast's body have the same rate of spin (i.e., angular velocity), the farther away from the bar they are, the longer the radius of gyration is, and the faster the body parts will move.

Figure 6.2 shows that in a front giant, the gymnast's feet (a) move around faster than his hips (b), which in turn move faster than his shoulders. The gymnast's fingers gripping the bar (c) move slowest of all, yet his fingers have the same angular velocity as all other parts of his body.

Can you determine, by looking at the figure carefully, which factors influence how fast the

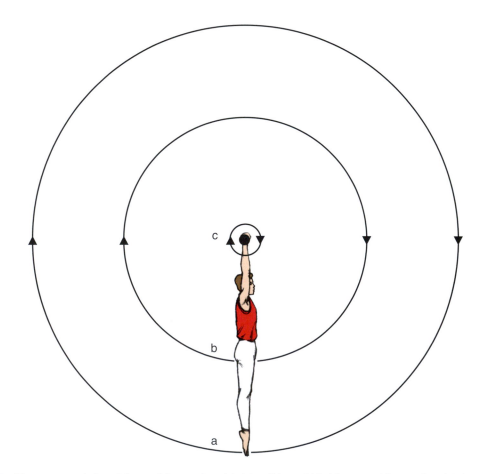

FIGURE 6.2   The gymnast's feet (a) travel faster than his hips (b), and his hips travel faster than his hands (c).

gymnast's feet will travel? Their speed depends on how many revolutions per second the gymnast rotates (which is the gymnast's angular velocity) and how far the gymnast's feet are from the axis of rotation (the words *how far* refer to their radius, or distance, from the axis of rotation). The axis of rotation is the high bar itself.

Let's take what we've learned from the high-bar example and apply it to a golf scenario. Imagine that you're getting a lesson from your club pro.

- The pro tells you, "If you want to drive the ball farther you'll need to produce more club-head speed."

- Now you can think to yourself, "OK, for more club-head speed I can increase the angular velocity of my club by swinging it faster through its arc, or I can hold higher up on the grip of the club to make the club length longer."

- "By gripping higher on the club I increase the distance (or radius of gyration) from my axis of rotation to the club head. I can even use a longer club like a driver, rather than a five iron. Any of these changes will increase the speed of the club head" (see figure 6.3).

Long-drive competitions in which athletes swing extremely long clubs at phenomenal angular velocities produce drives of more than 350 m (382 yd). Long-drive competitions don't require extreme accuracy, but they certainly demonstrate

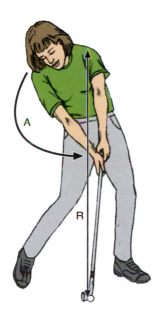

FIGURE 6.3 In a golf swing, angular velocity (A) multiplied by the radius (R) determines the speed of the club head.

what's required to hit a golf ball a long way (Read et al. 2013).

You can apply the reasoning that you used during your golf lesson to any sport that uses bats, rackets, or clubs. To hit a ball as hard as possible, batters should hold higher up toward the end of the grip to increase the radius and then swing as fast as they can. Swinging the bat through its arc as quickly as possible maximizes its angular velocity. Maximum angular velocity combined with the optimal radius gives the greatest speed to the striking portion of a club, bat, or racket.

Centripetal and Centrifugal Force

Whenever rotation happens, interplay occurs between inertia, centripetal forces, and centrifugal forces. These forces are present in the spin of a volleyball spike serve, a spiral pass by a football quarterback, and even a gentle putting stroke by a golfer. Rotation is a battle between inertia and centripetal and centrifugal forces. The inertia of an object when it is moving is expressed in its "desire" to travel in a straight line; remember that inertia is the resistance to changing the state of motion.

- Changing straight-line motion into circular motion requires a **centripetal force**. This force pulls (or pushes) an object toward the axis of rotation to make it follow a curved, or circular, pathway.

- When an athlete swings a baseball bat, he applies a centripetal force to make sure that the bat follows the arc of the swing.

- The inward pull of centripetal force produced by the athlete battles the inertial desire of the bat to travel in a straight line.

- The athlete will certainly feel that the bat is trying to pull outward away from his grip; therefore, batters lean back when swinging a bat to help counter these forces (Newton's action–reaction law).

- This outward pull is often called a **centrifugal force**; centrifugal means "pulling outward from the axis of rotation."

- *Centrifugal force* is a commonly used term, yet it is characterized by physicists as a phony or fictitious force. It is best to think of centrifugal force as inertia in disguise. Using the hammer throw as an example, the following description will explain why.

To understand how inertia, centrifugal forces, and centripetal forces are interrelated, let's have an

Olympic hammer thrower spin around and throw the hammer. Like anything else that has mass, a hammer has no wish to move to begin with and, second, no desire to follow a circular path. The inertia of the hammer makes it resist movement. A competitive hammer for men weighs 7.3 kg (16 lb) and for females 4 kg (8 lb 12 oz), so these implements have plenty of mass and, consequently, a lot of inertia. If an athlete applies sufficient muscular force to get the heavy ball of a hammer to follow a circular pathway, then its inertia will be expressed by its desire to travel in a straight line and not around a circular pathway.

When athletes make a hammer follow a circular pathway, they apply a centripetal force by pulling inward on the hammer at every instant the hammer goes around. As the athletes pull inward on the hammer, they experience an outward centrifugal force. Because the athletes provide the axis of rotation around which the hammer travels, they certainly feel that the hammer is trying to drag them outward. In reality, this outward pull is the hammer wanting to travel along a straight line or, more precisely, to fly off at a tangent to the circle that it is being forced to follow. The more massive the hammer is, or the faster the athlete makes it move (Judge et al. 2016), the greater its resistance (inertia) is to following a circular pathway and the greater its desire is to fly off in a straight line (see figure 6.4).

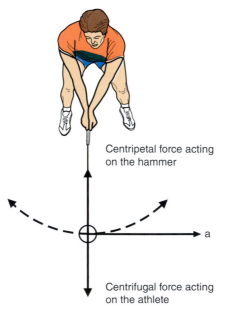

FIGURE 6.4 Centripetal and centrifugal force in the hammer throw. The hammer wants to travel in the direction of arrow a.

Immediately upon release, the hammer gets its chance to quit going around in a big circle and instead fly away along a straight-line pathway. And if it were not for gravity and air resistance, it would do just that, following the trajectory that the athlete gave it at the instant of release. Throwers no longer apply a centripetal force to the hammer after they've released it. The distance the hammer travels through the air depends on the velocity and trajectory of the hammer ball at the instant of release—not on centripetal force and certainly not on centrifugal force.

The amount of centripetal force that hammer throwers produce depends on what occurs.

- For example, if a thrower doubles his angular velocity by spinning around twice as fast, then he must increase his inward pull on the hammer ball fourfold. Why? Because the hammer goes around twice as fast and tries to follow a straight-line pathway at every instant with twice the force.

- In other words, the increase in the athlete's rate of spin increases to the second power, or squares, the demand for centripetal force.

- To double the rate of spin (i.e., angular velocity), the athlete has to pull inward (i.e., increase the centripetal force) four times as much.

- To triple the rate of spin, the athlete has to increase the centripetal force nine times.

What happens if no change occurs in the thrower's angular velocity (i.e., rate of spin), but instead he spins around with a hammer twice as heavy? In this case, the athlete must double his centripetal force because the extra mass of the hammer tries to travel off in a straight line at every instant with twice the force. Athletes who spin around with a more massive hammer or who attempt to spin faster (i.e., increase their angular velocity) must be prepared to increase their centripetal force by flexing their legs, leaning backward, and pushing against the earth. Leaning backward and pushing with the legs against the earth causes the earth to push against them, and this action counteracts the extra pull of the hammer (see figure 6.5). Failure to carry out these actions can throw athletes off balance, often with disastrous effects.

Keep in mind that the battle between inertia and centripetal and centrifugal forces occurs in all sport skills in which rotation occurs, not just in hammer throwing. For example, when speed skaters glide around the oval and lean into a curve, they are

FIGURE 6.5 As the hammer travels faster, the thrower increases centripetal force by leaning away (arrow a) from the pull of the hammer (arrow b).

obeying the same principles as the hammer throwers. By leaning into the curve, speed skaters push outward at an angle against the ice (i.e., against the earth). The earth pushes back (equal and opposite) and provides the inward push of a centripetal force. The more massive the skaters are, the faster they travel; the tighter the curve is, the more they must lean and push outward against the earth to get the earth to push them inward.

AT A GLANCE

What Happens When We Rotate

- *Angular velocity* is a term used to describe the rate of spin of an athlete or an object.
- Whenever rotation happens, interplay occurs between inertia, centripetal forces, and centrifugal forces.
- Changing straight-line motion into circular motion requires a centripetal force. This outward pull is often called a centrifugal force; centrifugal means "pulling outward from the axis of rotation."

The fact that a speed skater must push outward to get around the curve indicates that some of the athlete's leg thrust is used to push outward while the remainder is used for propulsion forward along the ice. After the athlete is out of the curve and into the straightaway, then all of the athlete's force can be used to drive toward the finish line. The situation is the same for a track sprinter rounding the curve in a 200 m sprint. Compared with running on a straightaway, where all the sprinter's effort can be spent driving toward the tape, running a curve requires the athlete to spend some precious energy thrusting outward to get around the curve.

Proof of this division of effort shows up in comparing the records for 220 yd on a straightaway and 200 m on a curve. (Keep in mind that 220 yd is 1.28 yd longer than 200 m.)

- In 1966 Tommie Smith ran the 220 yd on a straightaway in 19.5 s, and in 1968 he set the world record for 200 m at 19.83 s.
- Therefore, although the 220 yd was longer, the straight-line running was 0.33 s quicker.
- In 1979 Pietro Mennea of Italy set a world record for the 200 m around a curve at 19.72 s.
- This record was broken by Michael Johnson (United States) at the Atlanta Games in 1996 with a time of 19.32 s, and then at the 2008 Beijing Games when Usain Bolt lowered this time to 19.30 s. Can you imagine what times Bolt might have run on a straight?

Is there any way to help a sprinter negotiate a curve without expending precious energy pushing outward? Yes. Bank the curves of the track. A banked track pushes the athlete inward in the same way that the drum of a washing machine holds clothes to a circular pathway during the spin cycle. On tight indoor tracks, high banking allows the athletes to run flat out without fearing that their inertia will cause them to fly off the track and into the spectators. Outdoor tracks are not normally banked, so the tight inside lanes can be difficult to negotiate. Particularly affected are heavier sprinters, who have more inertia and must push outward more vigorously to get their extra body mass around the curve.

Rotary Inertia

Rotary inertia can be thought of as rotary resistance or rotary persistence. Physics texts commonly described it as "moment of inertia." All four terms mean the same thing. **Rotary inertia** is the tendency of all objects, or all athletes, to resist rotation initially and then to want to continue rotating after they have had the turning effect of torque applied against them and when a centripetal force is keeping them following a circular pathway. Rotary

APPLICATION TO SPORT

Traveling Around a Banked Track at High Speed

Athletes who race indoors are familiar with tracks that are banked on the curves. The banked track allows athletes to run flat out without having to reduce their speed as they negotiate the curves. Texas Motor Speedway in Fort Worth, Texas, is banked at 24°. The steep banking allows CART race cars to hurtle continuously around the track at full throttle. Drivers complained of dizziness, vertigo, and other problems with their vision and hearing after traveling at speeds in excess of 322 km/hr (200 mph) around the speedway. These physiological difficulties can be related to a battle that occurs between centripetal force and inertia. Centripetal force keeps the race car and driver following the curve of the track, whereas inertia wants the race car and driver to travel in a straight line. The outward inertial pull on the drivers as they rounded the track was estimated at 5 g, or five times their normal body weight.

Jeff Gross/Getty Images

inertia occurs in every situation in which athletes rotate, spin, or twist and in every situation in which bats, clubs, and other implements are swung. In short, rotary inertia exists in all sporting situations in which angular (i.e., rotary) motion occurs.

Rotary inertia is the rotary equivalent of linear inertia and therefore is related to Newton's first law of inertia. Massive linemen have great linear inertia. Their mass doesn't want to move, and if forced to move, it wants to travel in a straight line. If these linemen got on a merry-go-round (like those in your local playground), a lot of effort (i.e., torque) would be required to get the merry-go-round spinning. But once underway, the merry-go-round (with the added mass of the linemen) would want to continue spinning.

If the teammates of the linemen got on as well, they would increase the merry-go-round's rotary inertia even further. The merry-go-round with all its riders would act like a giant flywheel—similar to the ones you see on old steam engines driving threshing machines at agricultural fairs. With all these heavyweight riders, a lot of torque would be needed to get the merry-go-round spinning, but after it was spinning it would want to keep spinning. The following two important factors determine how much inertia a rotating object will have:

1. *The mass of the object.* The more massive an object is, the more resistance it puts up against being rotated. In addition, the more mass an object has, the greater the persistence it has in wanting to continue rotating after rotation is established. A heavy (i.e., more massive) baseball bat is more difficult to swing than a light one. It resists being accelerated through the swing more than the lighter bat does. After a batter has applied sufficient turning effect, or torque, to get a bat moving, a heavy bat wants to continue the swing more than a lighter one does. The heavier the bat is, the stronger the athlete must be to get it moving and to control and stop it after this motion has been initiated.

2. *The radial distribution of mass.* The phrase *radial distribution of mass* refers to how the mass of an object is distributed (i.e., positioned) relative to the axis around which it's spinning. Two extremes in the distribution of mass have to do with whether the object's mass is far from the axis of rotation or close to the axis of rotation.

Here's an example of these two factors in a sport situation. Imagine that you are given two golf clubs—club A and club B. The two clubs are alike in length and shape, and on a scale their mass is exactly the same.

- Club A is like any other golf club you would buy in a sporting goods store.

- But as soon as you pick up club B, you sense that it differs tremendously from club A. Except for a small amount of mass in the shaft, all of its mass has been concentrated in the club head.

- Club B will therefore have more rotary inertia than club A because almost all of its mass is way out in the club head.

- Compared with a swing with club A, a swing with club B will be more difficult to initiate. It will also be more difficult to stop after the swing is underway.

Let's take club B, with most of its mass in the head, and start repositioning its mass by shifting it up the shaft of the club toward the grip. We'll leave the club head the same shape but hollow it out so that it has hardly any mass. As we progressively move the mass of the club from the head toward the grip, its rotary inertia (i.e., rotary resistance) is gradually reduced so that swinging it becomes easier and likewise it becomes easier to maneuver during the swing. Finally, when almost all of its mass is in the grip, the club will feel like a fencing foil, which has most of its mass in the handle. You'll feel that you can maneuver this golf club quite easily, but you'll also know that a club designed in this fashion will never hit a golf ball very far.

How Differences in Distribution of Mass Affect Rotary Inertia

We have just seen that the rotary inertia of any object, whether it is a golf club or an athlete, depends on

- how much mass it has (the more mass it has, the more rotary inertia it has) and

- how its mass is positioned relative to its axis of rotation (the farther the mass is from the axis of rotation, the greater the radius of gyration is and therefore the greater the rotary inertia is).

How does the positioning of the mass of an object relative to its axis of rotation affect its resistance to being made to rotate and, after it is rotating, being made to stop rotating? The answer to this question illustrates the huge difference between the inertia of an object traveling in a straight line and that of an object that is spinning. Objects traveling in a straight line increase their inertia in direct proportion to their mass. More mass equals more inertia. But with a rotating object, whenever

its mass is moved closer to the axis, or farther from the axis, a dramatic change occurs in the object's resistance to rotation. Here's an example of how this phenomenon works.

Imagine that you have a ball with every particle of its mass concentrated at its center (of course, all the mass of a ball cannot possibly be at its center, but for this example let's imagine that it is).

- In figure 6.6a we will discount gravity and air resistance. The ball is on a string and rotating around an axis at 1 rev/s. The distance of the ball from the axis is 2 units.

- In figure 6.6b this distance is suddenly reduced to half of the original. This reduction in radius causes the rotary inertia of the ball to be reduced fourfold (i.e., to a quarter of its original). Because of this huge reduction in rotary inertia, the ball finds it easier to rotate, so it speeds up (i.e., increases its angular velocity) proportionally from 1 rev/s to 4 rev/s.

- In figure 6.6c the distance of the ball from the axis has been doubled from 2 units to 4. The rotary inertia of the ball is now

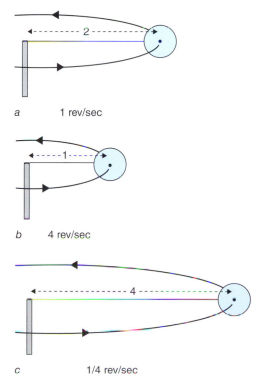

a   1 rev/sec

b   4 rev/sec

c   1/4 rev/sec

FIGURE 6.6 Reducing the radius from (a) 2 units to (b) 1 unit reduces rotary inertia fourfold. Doubling the radius from (a) 2 units to (c) 4 units increases rotary inertia fourfold (note that air resistance and gravity have been discounted).

increased fourfold. The ball finds it more difficult to rotate, so it slows to 1/4 rev/s.

This example tells us that the rotary inertia of a spinning object is proportional to the square of the radius. Halve the radius, and the rotary inertia is reduced to one-quarter of the original (1/2 × 1/2). Double the radius, and the rotary inertia increases to four times the original (2 × 2). What is the effect of changing the mass of the rotating object? The answer is that it is directly proportional. Doubling the mass of the object doubles its rotary inertia. Halving it halves its rotary inertia.

In no situation in sport does every little bit of an object or an athlete shift the same distance toward or away from its axis. You cannot move every particle of a baseball bat to one end because nothing would be left to make up the remaining part of the bat. Nor can you shift every particle of the bat halfway along its length because that would leave nothing at the ends. It's the same with athletes.

But athletes and inanimate objects such as baseball bats differ significantly! Athletes can change their shape at will. They can tuck their bodies up tight, or they can extend them. By carrying out these maneuvers, they can pull their body mass in close to their axis of rotation or push their body mass out from their axis of rotation as far as possible. Let's see how this occurs in sport situations.

Manipulating Rotary Inertia in Sport

When springboard and tower divers somersault in the air and move from an extended body position to a tuck, they flex their torsos, legs, and arms. Some parts of their bodies (i.e., their arms and legs) shift a large distance toward their transverse axis (hip to hip). Other parts, such as their heads, move a short distance. Nevertheless, a dramatic difference in rotary inertia occurs between an extended and a tucked body position. A diver's legs and arms are relatively heavy and have a lot of body mass. Moving them a large distance toward the diver's axis greatly reduces the resistance of the diver's body against rotation.

Divers in flight rotate slowly when they assume an extended body position. If they pull their bodies into a tight tuck, they rotate much faster, as shown in figure 6.7. The more body mass they pull toward their axis of rotation (which always passes through their center of gravity), the faster they spin. Divers with lean bodies and great **flexibility** can spin much faster when they pull into a tuck. Huge, muscular athletes have a tough time pulling themselves in as tight as lean athletes. Their excess body mass gets in the way. The technique used for multiple somersaults is similar to that used by trapeze artists. Trapeze artists perform quadruple somersaults by gripping their shins and pulling

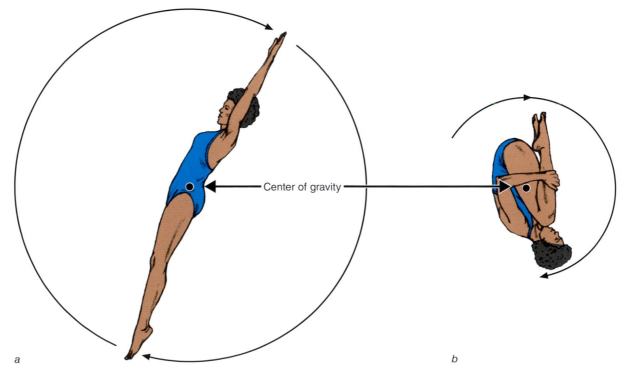

a   Center of gravity   *b*

FIGURE 6.7 Angular velocity increases as the diver (*a*) pulls her body in close (*b*) to her axis of rotation, which, in flight, passes through her center of gravity.

their knees up as high as possible toward their shoulders. The more compressed and compact they are, the faster they spin.

Athletes vary their rotary inertia in many events, not just in sports like diving. For example, in sprinting 100 m, a sprinter extends his legs to the rear of his body and flexes them when they come forward for the next stride. Likewise, when he drives his legs forward, the thigh is elevated and the leg is flexed at the knee. Flexing at the knee brings the mass of the sprinter's leg closer to the hip joint (the axis around which the leg rotates). This action reduces the leg's rotary inertia, as shown in figure 6.8. If the rotary inertia of the leg is reduced, the task of moving the leg forward becomes much easier for the muscles involved.

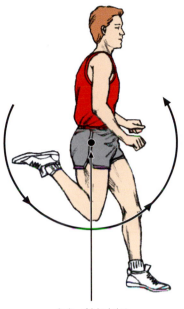

Axis of hip joint

FIGURE 6.8 Rotary inertia of the sprinter's leg is reduced when it is flexed.

AT A GLANCE

Rotary Inertia

- Rotary inertia is the tendency of all objects, or all athletes, to maintain their current state of rotation (either at rest or in motion).
- The more massive an object is, the more resistant it is to changing its state of rotation.
- The distribution (i.e., position) of mass, relative to the axis around which it's spinning, also influences rotary inertia.

Radius of Gyration

The rate or amount of rotation for any object (human or implement) can be influenced by its **radius of gyration**. As shown in figure 6.9, a gymnast is performing a front giant, which is one

APPLICATION TO SPORT

Technology and the Hammer Throw

The men's hammer that is thrown in track and field weighs 7.3 kg (16 lb) and measures 1.2 m (4 ft) from the handle to the farthest point on the surface of the ball. Until the rules were changed, technicians realized that if they used extremely dense tungsten, they could put almost all of the hammer's 7.3 kg in the distal portion of the ball. This design shifted the hammer's center of gravity farther away from the thrower and made the ball about the size of a baseball, which reduced its air resistance during flight. The handle was then made of extremely light titanium so that virtually no weight was close to the thrower. Unfortunately, an unexpected problem developed—hammers with 7.3 kg concentrated in a sphere as small as a baseball buried themselves so deep in the turf that athletes and officials had a hard time pulling them out! Rules now outlaw these types of hammers.

Adam Nurkiewicz/Getty Images for European Athletics

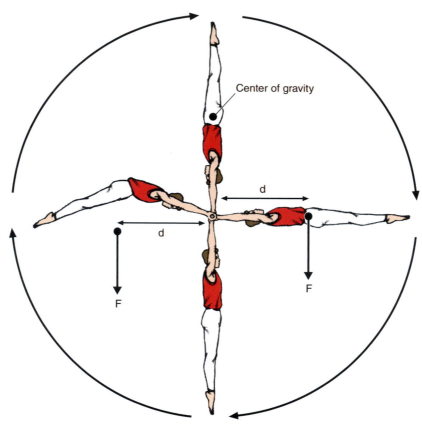

d = Distance of center of gravity from axis

FIGURE 6.9 Radius of gyration in a front giant around the high bar.

360° revolution around the high bar. The radius of gyration is the distance from the axis of rotation (the point at which the object is rotating about) and the center of gravity (the balance point) for the segment that is rotating.

The gymnast performing the front giant rotation has extended the arms and legs to make the body

AT A GLANCE

Angular Motion Fundamentals

- The fundamental components for anything to rotate are an axis of rotation (fulcrum or pivot point) and a lever attached to this reference point.
- The effect of this rotation is often measured from the object's (human or implement) center of gravity, that is, the balance point at which the weight force is equal on either side.
- Angular motion or rotation occurs when the force (and the resistance) is applied at a distance from the axis of rotation.

as long as possible. This positioning in turn makes the radius of gyration as large as possible, so that when the gymnast is in the horizontal position, he will (because of gravity) have the largest amount of torque for rotation. We'll talk more about the radius of gyration in the following sections. The greatest turning effect (i.e., torque) applied by gravity occurs as the gymnast passes through this horizontal position. Conversely, no torque is produced by gravity when the gymnast is directly above or directly below the bar.

Applied Angular Motion: How an Athlete Initiates Rotation

As discussed earlier, rotation occurs when the force (and the resistance) is applied a distance from the axis of rotation. This offset creates the turning effect of torque and happens whenever a force is applied at a distance from an axis. If we increase the distance or the force that an athlete applies, the torque turning effect becomes greater. Therefore, a longer limb, or a stronger muscle force, will make the segment under load move faster. The same principle applies for an object as for an athlete, and the application of torque can make objects and athletes rotate.

Rotation of an Object

In ball games, such as tennis or volleyball, the server wants to spin the ball on some occasions and avoid spinning it on others. If volleyball players wish to serve a floater, they make sure that the force they apply to the ball passes directly through the ball's center of gravity, which in flight is the ball's axis of rotation (Quiroga et al. 2010). When this happens, the ball floats across the net without spinning (figure 6.10).

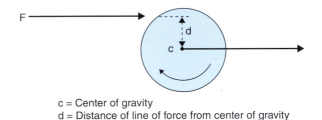

c = Center of gravity
d = Distance of line of force from center of gravity

FIGURE 6.11 Application of topspin to a ball.

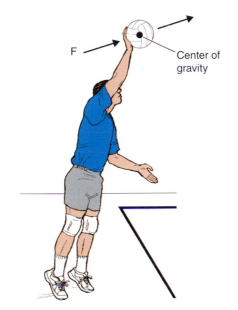

FIGURE 6.10 No spin is imparted to a volleyball when force is directed through its center of gravity.

Adapted from K. Luttgens and K. Wells, *Kinesiology: Scientific Basis of Human Motion*, 8th ed. (New York: Times Mirror Higher Education Group, 1992), 341.

What must happen if servers want to spin the ball? In this case, volleyball players apply the force from their hand at some distance from the ball's center of gravity. Increasing this distance magnifies the spin. When this happens the ball receives topspin, as shown in figure 6.11, which results in a downward arc in flight.

In golf, a chip shot requires golfers to apply force well below the ball's center of gravity.

- The angle of the club face and the stroke technique produce backspin, which causes the ball to lift in flight.

- The amount of backspin influences how much the ball will rise.

- Depending on the surface on which it lands, the ball can stop dead or, if shot to the far side of the green, roll back toward the pin (Richardson et al. 2017).

The amount of spin given to any ball depends on how much force is applied and how far it is applied from the ball's center of gravity. The greater the force is and the larger the distance is from the center of gravity, the greater the torque and the spin are. If you want to apply maximum spin to a football, you grip it closer to the middle where the ball is fattest, as shown in figure 6.12*a*.

Here, the force applied by the fingers to the circumference of the ball is farthest from the ball's center of gravity and its long (i.e., longitudinal) axis. In this way maximum torque is applied, which gives the ball the greatest amount of spin. The drawback with this position is that less force is put into throwing the ball for distance. On the other hand, the passer who wants to apply less spin and concentrate on applying maximum force to the ball grips it closer to the end, as shown in figure 6.12*b*. Most quarterbacks compromise by gripping the ball halfway between these two positions.

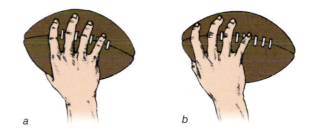

FIGURE 6.12 Varying spin and directional force of a football. (*a*) Gripping the middle of the ball provides maximal spin and less force for distance; (*b*) gripping the end of the ball provides less spin and maximal force for distance.

How Athletes Make Themselves Rotate

Athletes who want to rotate in the air employ the same mechanical principles to themselves as those used to apply spin to a volleyball. For example, a trampolinist who bounces with his center of gravity directly above the upward thrust of the trampoline bed rises vertically without rotating. The thrust of the trampoline pushes directly upward through his center of gravity (Yeadon and Hiley 2017). Like the volleyball, the trampolinist moves vertically, and without rotating, in the same direction as the thrust of the trampoline (see figure 6.13a).

The key difference between the rotation of an athlete and the rotation of an object is the speed of rotation. Because a ball can rotate and spin significantly faster than a human, the ball invokes the Magnus effect, in which the spin of the ball can alter the flight path (discussed in more detail in chapter 10). To date no human has rotated fast enough to generate a curved path because of pressure differential, but the human's asymmetrical shape allows us to distort the flight path.

If the trampolinist positions his center of gravity so that it is no longer directly above the upward thrust of the trampoline bed, he has a tendency to rotate. The more the athlete shifts his center of gravity out of line with the thrust of the trampoline (e.g., by leaning forward or backward), the greater the turning effect (i.e., torque) applied to his body is and the greater the tendency is for rotation to occur (see figure 6.13b).

How to Measure Angular Velocity During Cycling

In this section we will measure angular velocity during cycling. To measure angular velocity, we use a procedure and technology similar to those used to measure linear movement—with a common video camera. Using this objective method, we can quantify the amount of rotation and therefore determine the influence of any change in athletic performance or determine why one athlete is running faster than another.

The key difference between linear and angular velocity is the magnitude. If elite athletes can run 100 m in, say, 10 s, they have an average linear velocity of 10 m/s. For a golf club swing, the average angular velocity at impact for Professional Golf Association players is around 50 m/s and for low-handicap players 40 m/s, which is four to five times greater in magnitude than the velocity of the sprinter.

Angular velocity (which is written as the Greek symbol omega, ω) can be measured in a number of ways, but regardless of the method used there must be one common theme—the measurement of change in position and the measurement of the change in time.

This principle is the same as that used with linear velocity measurement. Because of the faster movements that occur with angular velocity, a more effective process for measuring angular velocity is by using video. The captured video image allows the movement process to be slowed down so that the actual reference points can be determined.

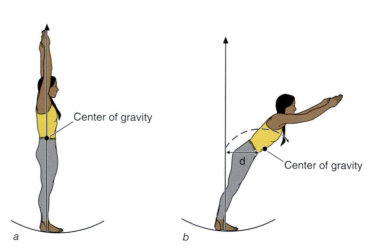

FIGURE 6.13   (a) A trampolinist rises vertically when the vertical thrust of the trampoline is through the center of gravity. (b) When the thrust from the trampoline does not pass through the gymnast's center of gravity, the torque produced causes the gymnast to rotate.

Adapted from H. Braecklein, *Trampolinturnen II* (Wiebelsheim, Germany: Limpert Verlag, 1974), 29.

The activity, as described next, determines the relationship between torque and the cadence of bicycle crank movement. We can measure the average cadence of pedaling by counting the number of pedal revolutions and recording the time in which they occur.

- For example, if 25 revolutions of the pedal occur in a 30 s period, the cycle cadence is 50 rev/min.

- We can calculate the instantaneous angular velocity from viewing the captured video in a frame-by-frame process, measuring the change in angular position (which is written as the Greek symbol theta, θ), with respect to time, and tracking this on a form such as the one shown in table 6.1.

- Making these measurements requires a stationary bicycle (preferably with a load device, such as a Monark cycle), stopwatches, a video camera, a TV or computer screen, tracing paper, and a protractor or goniometer. These are the steps for measuring angular velocity:

1. Set up the video camera perpendicular to the bicycle (follow the setup guidelines in the previous chapter).

2. Measure the lever arm of the pedals [____ (m)].

3. Adjust the load on the cycle to minimum, have the subject pedal at maximal effort for 1 min, and note the cadence for that load.

4. Apply a load just to the point where the subject cannot move the pedals; this is the maximum force.

5. Divide the maximum force into three equal loads: (a) maximal, (b) midway,

and (c) minimal. For each incremental load, have the subject pedal at maximum effort for 1 min. Note the cadence for that load.

6. When playing back the video, place the tracing paper on the screen and mark the axis of rotation and the center of the pedal. Then move one frame forward on the video. Mark the new position of the pedal. Continue this process for five consecutive frames.

7. Measure the change in angle (degrees) that the pedal goes through from one frame to the next; this is the angle the pedal has moved through in the frame time (with the NTSC video system this will be 30 hertz, or 0.033 s; with a PAL system this will be 25 hertz, or 0.04 s).

8. The angular velocity (θ) is change in angle divided by the change in time.

Perfect Angular Velocity in Sport?

In the optimum sport performance, the movement of the athlete is smooth and effortless. If we plot the angular velocity against time for any movement, each segment always follows a similar curved (or parabolic) pattern. This circumstance reflects the fact that the angular velocity is zero at the start of the movement and returns to zero at the end of the movement and that somewhere in between (usually at midway) is the fastest angular velocity.

Let's take the example of kicking a football. The objective is to generate the maximum distance or velocity (or both) of the ball. For this to happen, the maximum angular velocity of the most distal segment must occur just as the foot makes contact with the ball. Athletes can achieve this in sport by collectively generating the maximum angular velocity

TABLE 6.1 Spreadsheet to Calculate Angular Rotation and Velocity for Different Loads

	Time (s)	Load 1 Angle θ (deg)	Load 1 Angular velocity ω (deg/s)	Load 2 Angle θ (deg)	Load 2 Angular velocity ω (deg/s)	Load 3 Angle θ (deg)	Load 3 Angular velocity ω (deg/s)
1							
2							
3							
4							
5							

APPLICATION TO SPORT

The Perfect Golf Swing

People often ask what makes elite golfers like Dustin Johnson and Annika Sorenstam so good. Do they have a perfect male or female golf swing? To answer the question, recognize that no "perfect" golf swing will fit each body type. Each person can try to generate a perfect swing for his body, at that point in time. As golfers mature, the range of movement and strength of their muscles may necessitate a different "perfect" swing.

We in sport mechanics can tell if the swing is perfect by assessing the kinetic link, or the body and limb coordination. Novices in the sport generally keep their knees fixed and have little trunk rotation. The swing involves rotating only the arms with the elbows and wrists locked. At the other end of the golf swing spectrum, the perfect swing starts with rotation at the core or trunk. As the trunk rotation reaches a maximum, the upper torso or shoulders begin to rotate (but because of the previous movement of the trunk, the shoulders already possess some rotation before that joint begins to move). The perfect swing then has the elbows commence extension, just at the peak of shoulder rotation. This is followed by wrist extension, all

Dan Mullan/Getty Images

occurring the instant before the club head makes contact with the ball. With this series of movements joined together, the perfect kinetic link is generated.

---

in each of the preceding movements, starting at the most proximal joint.

A perfect example of this collective angular velocity is shown in figure 6.14. At the release

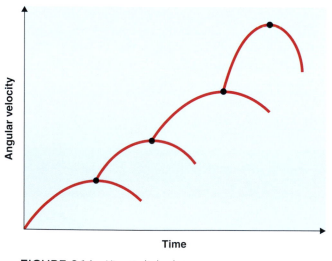

FIGURE 6.14   Kinetic link plot.

Adapted from E. Kreighbaum and K. Barthels, *Biomechanics: A Qualitative Approach for Studying Human Movement*, 3rd ed. (Upper Saddle River, NJ: Pearson Education, 1990), 343.

point, the maximum possible accumulated angular velocity is generated. When looking at this figure, we start at the bottom left corner. The first curve is the plot of the inertial velocity of the thigh, and the maximum angular velocity occurs approximately midway through this path. At that point in time the next distal segment, that is, the shank (or shin), should commence its swing. If the timing is perfect, the shank has the maximum possible starting angular velocity, so the cumulative angular velocity of the shank is the highest possible. If the same perfect timing occurs and the next distal segment, the ankle, starts its angular rotation at the maximum angular velocity of the shank, the cumulative angular velocity is the highest possible.

This process, known as the **kinetic link principle**, is a key ingredient for enhancing sport performance. Novice athletes commence the movement of the next distal segment either before or after achievement of the maximum angular velocity of the current segment, so the net cumulative angular velocity is lower. This kinetic link principle of angular movement is applicable to all sports that involve rotation, such as baseball, golf, and basketball.

SUMMARY

- All rotational, turning, spinning, and swinging motions are forms of angular motion. Angular motion implies that an object or athlete rotates around an axis.

- Motion in an athlete's body is predominantly rotational. Muscles pull on bones, and bones rotate at the joints.

- Angular velocity is synonymous with rate of spin. It refers to the angle, degrees, or revolutions completed in a particular period (e.g., 1 s) in a specific direction (e.g., clockwise or counterclockwise).

- All objects that rotate or swing have an inward pulling or pushing force, called a centripetal force, that acts toward the axis of rotation. Centripetal force counteracts the inertial desire of objects to travel in a straight line.

- When athletes or objects rotate, a centripetal force exists. In response they will experience the outward pull of what is commonly called centrifugal force. Centrifugal force is caused by the inertial desire of whatever is rotating to travel in a straight line and not in a circle. Centrifugal force is often considered a fictitious or phony force.

- The inertia of all objects makes them resist rotation. When forced to rotate, however, an object's inertia is expressed by wanting to continue rotating.

- Rotary inertia varies according to the mass of a spinning object and the way its mass is distributed. The greater the distance that mass is spread out from its axis of rotation, the greater the rotary inertia. The more compressed that mass is relative to its axis of rotation, the greater the reduction of rotary inertia.

KEY TERMS

angular motion	frontal axis
angular velocity	kinetic link principle
balance point	longitudinal axis
centrifugal force	radius of gyration
centripetal force	rotary inertia
flexibility	transverse axis

REFERENCES

Hiley, M.J., M.I. Jackson, and M.R. Yeadon. 2015. "Optimal Technique for Maximal Forward Rotating Vaults in Men's Gymnastics." *Human Movement Science* 42:117-31.

Judge, L.W., M. Judge, D.M. Bellar, I. Hunter, D.L. Hoover, and R. Broome. 2016. "The Integration of Sport Science and Coaching: A Case Study of an American Junior Record Holder in the Hammer Throw." *International Journal of Sports Science and Coaching* 11 (3): 422-35.

Quiroga, M.E., J.M. García-Manso, D. Rodríguez-Ruiz, S. Sarmiento, Y. De Saa, and M.P. Moreno. 2010. "Relation Between In-Game Role and Service Characteristics in Elite Women's Volleyball." *Journal of Strength and Conditioning Research* 24 (9): 2316-21.

Read, P. J., S.C. Miller, and A.N. Turner. 2013. "The Effects of Postactivation Potentiation on Golf Club Head Speed." *Journal of Strength and Conditioning Research* 27 (6): 1579-82.

Richardson, A.K., A.C.S. Mitchell, and G. Hughes. 2017. "The Effect of Dimple Error on the Horizontal Launch Angle and Side Spin of the Golf Ball During Putting." *Journal of Sports Sciences* 35 (3): 224-30.

Yeadon, M.R., and M.J. Hiley. 2017. "Twist Limits for Late Twisting Double Somersaults on Trampoline." *Journal of Biomechanics* 58:174-78.

 Visit the web resource for review questions and practical activities for the chapter.

7

Angular Kinetics in Sport

Chris Hyde/Getty Images

As the name suggests, **angular kinetics** is all about rotation and the forces experienced when rotating. A key influence on rotation is the influence of **gravity**. Gravity can cause and initiate rotation or limit rotation. For many activities in daily living or sport we need to overcome the force of gravity. When an object or human is rotating, it typically turns about a pivot point, hence the term angular. This rotation can be at an external end point (like a gymnast on a bar), or from an internal balance point, such as your center of gravity. In sport we aim to vary these fundamental physical concepts (gravity and rotating) to alter the timing and rate of angular rotation, which ultimately produces the desired sport performance. To understand angular kinetics we initially look at the fundamental principle of angular kinetics, gravity. We then look at shifting the center of gravity to manipulate the flight path, which leads into the conservation and **transfer of angular momentum**.

Gravity: Fundamental Principle of Angular Kinetics

Gravity is an invisible force that is all around us on earth, but because we can't see it directly, it is sometimes a difficult concept to comprehend. Gravitational forces vary slightly at various places on earth. How do these differences affect performance in sport? As an example, let's look at some of the venues where the Olympic Games have been held.

Athletes experienced slightly less gravitational pull at the 1968 Olympics in Mexico City, which is at a higher altitude and closer to the equator, than they did at the 1952 Olympics in Helsinki or the 1980 Games in Moscow, both of which are in northern latitudes and closer to sea level. Considering gravity by itself and not air resistance, Peter Brancazio, an avid sport fan and physics professor at Brooklyn College, calculated that a 20 m (66 ft)

shot put in Oslo, Norway (latitude 60° N) would travel

- 25 mm (1 in.) farther in Montreal, Canada (45° N),
- 50 mm (2 in.) farther in Cairo, Egypt (30° N), and
- 75 mm (3 in.) farther in Caracas, Venezuela (10° N).

Furthermore, a 100 m (328 ft) javelin throw in Moscow (56° N) would travel 90 m (295 ft) in Lima, Peru (12° S).

Of greater importance than the slight reduction in the pull of gravity is the so-called thin air present at high altitudes. Although air contains the same proportions of oxygen (21%), nitrogen (78%), and other gases (1%) at high altitudes as it does at sea level, less of each is found in a similar volume of air the higher up in altitude we go. This characteristic greatly affected athletes who competed in the 1968 Olympic Games at Mexico City, which is 2,240 m (7,350 ft) above sea level.

At Mexico City, athletes had to breathe more vigorously and more often to get the oxygen they needed. The thin air caused a serious problem for athletes in endurance events, but it assisted athletes in short sprints because they ran on their bodies' stored energy supplies. When Bob Beamon set his world record in the long jump in Mexico City, he benefited (as did all the other long jumpers) from a slight reduction in gravity, reduced air resistance from less dense air (as he sprinted down the runway), and the fact that his approach was a short sprint and not a distance run.

After standing for many years, Bob Beamon's record was beaten by Mike Powell in the 1991 World Track and Field Championships in Tokyo. Tokyo is close to sea level, at a much lower altitude than Mexico City. So if you consider only the differences in atmospheric conditions, you can assume that Mike Powell's jump in Tokyo would have produced a greater distance had he performed it in Mexico City. (Note that the track and the approach in Tokyo used an ultraspringy artificial surface that was subsequently banned from further use. It was concluded that this surface assisted the sprinters and was likely of assistance to long jumpers as well.)

Acceleration Due to Gravity

When a pole-vaulter drops from above the bar, gravity accelerates the athlete toward the pit. If the athlete clears the bar at 6.1 m (20 ft) rather than at 4.6 m (15 ft), the extra distance gives the earth more time to accelerate the athlete on the way down. Dropping from 6.1 m, a pole-vaulter hits the pit at a greater velocity than when clearing 4.6 m.

The pole-vaulter's acceleration toward the earth is similarly experienced by a tower diver heading toward the water. Because of earth's gravitational pull, a diver continuously accelerates during the fall to the water. Figure 7.1 discounts air resistance

and shows an athlete stepping off and dropping from a height of 78 m (256 ft).

- After 1 s of fall, the athlete is traveling at a velocity of 9.8 m/s (32 ft/s), or 35 km/h (21.8 mph).
- After 2 s, the athlete's velocity has reached 19.6 m/s (64 ft/s), or 70 km/h (43.6 mph).
- At the 3 s mark, the athlete has reached 29.4 m/s (96 ft/s), or 105 km/h (65.4 mph).
- Finally, at the 4 s mark, the athlete's velocity has increased to an incredible 39.2 m/s (128 ft/s), or 140 km/h (87.2 mph).

As noted earlier, because of the regular addition of a speed of 9.8 m/s (32 ft/s) for every second, and discounting air resistance, we say that the force of gravity uniformly accelerates a falling athlete, such as a diver, trampolinist, or pole-vaulter, by a velocity of 9.8 m/s (32 ft/s) for every second of fall, or 9.8 m/s² (32 ft/s²).

How can we get some idea of the effect of **gravitational acceleration**? Look again at figure 7.1 and check the distance that the athlete covers with each second of fall.

- After 1 s, the athlete has fallen 4.8 m (16 ft). This distance is not too far, but gravity has had only 1 s to accelerate the athlete downward from board level.
- By the 2 s mark, the athlete has fallen a distance of 19.6 m (64 ft).
- At the 3 s mark, the athlete has reached 43.9 m (144 ft).
- Finally, by 4 s, the distance covered is an amazing 78.0 m (256 ft).

The athlete accelerates downward at a uniform rate of 9.8 m/s² (32 ft/s²), so with each passing second, the athlete adds on an additional 35 km/h (21.8 mph). Because of the constant increase in velocity, the athlete covers increasingly large stretches of distance with each second that passes.

The phenomenal acceleration caused by the pull of gravity makes tower diving a risky sport. A standard tower is 10 m (just under 33 ft) from the surface of the water. Tower divers take between 1.50 and 1.75 s to reach the surface of the water, and they are traveling at close to 61 km/h (38 mph) when they enter! Water is hard when divers hit it at that speed. Divers often wear wraps to provide additional support to their wrists, because they use a palm-first entry rather than one with the fingertips leading (Haase 2017). A palm-first position with

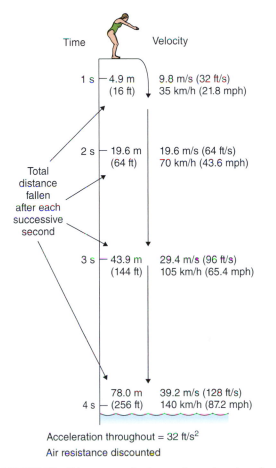

Time   **Velocity**

1 s — 4.9 m   9.8 m/s (32 ft/s)
      (16 ft)   35 km/h (21.8 mph)

2 s — 19.6 m   19.6 m/s (64 ft/s)
      (64 ft)   70 km/h (43.6 mph)

Total
distance
fallen
after each
successive
second

3 s — 43.9 m   29.4 m/s (96 ft/s)
      (144 ft)   105 km/h (65.4 mph)

        78.0 m   39.2 m/s (128 ft/s)
4 s —   (256 ft)   140 km/h (87.2 mph)

Acceleration throughout = 32 ft/s²
Air resistance discounted

FIGURE 7.1   Distance, velocity, and acceleration due to gravity (air resistance has been discounted).

the hands bent back at the wrists helps produce a bubbling, splashless entry (i.e., a "rip entry").

Does the uniform acceleration of springboard and tower divers mean that a parachutist in a free fall from several thousand meters (or feet) up accelerates continuously toward the earth? No, because air resistance in the denser atmosphere close to the earth's surface increases to a point where a parachutist in a free fall reaches a constant (or terminal) velocity of about 200 km/h (125 mph). In this situation, the pull of gravity is counterbalanced by the resistance generated by the atmosphere. Variations in terminal velocity can occur depending on the body position used by the parachutist during the free fall. The commonly used belly-down, spread-eagle position produces a much slower terminal velocity than a headfirst position with legs together and arms by the sides. Differences in the types of parachutes also cause variations in the speed with which the parachutist returns to earth. Modern wing-shaped, ram-air parachutes are designed to provide greater maneuverability and a softer landing than the older half-sphere parachutes that were commonly used during World War II.

An Athlete's Center of Gravity

As already discussed, gravity pulls everything that has mass toward the center of the earth. Just as we can measure the velocity of the whole rigid body of athletes when they are running, we can measure the individual components of the arm or leg, for example.

- If we collectively add up all the parts of the object (human or implement), we can find a **balance point**, the point at which the weight force is equal on either side.

- This balance, or center point, is the object's **center of gravity**.

- Note, however, that this point is theoretical and is used only for calculations to describe the movement of an object.

For example, let's look at a wooden baseball bat. If we assume that each particle of the wood in the bat has the same mass, the earth's gravitational attraction pulls on each of these particles with the same force. If you join all these "pulls" together and combine them into a single force, the place where this force is concentrated would be the bat's center of gravity, which—because of the extra mass at the proximal end of the bat—would be around the bottom third of the length of the bat. This location,

often called the sweet spot, is the theoretical, but important, center of gravity location for calculating any movement of this object.

Here's another example. Imagine a wooden ruler used for measuring short distances. You can balance this ruler like a seesaw on your fingertip at a point halfway in from the ends and halfway in from the sides. Your fingertip supports the ruler directly below its center of gravity, and equal amounts of the ruler's mass are found lengthwise and widthwise from the ruler's center of gravity. What would happen if we attached a small piece of lead to one end of the ruler? To balance the ruler, you need to move your supporting finger toward the end of the ruler where the lead is attached. The lead plus a smaller section of the ruler now balances a longer section on the opposing side of your supporting finger. The balance point has moved because the lead has a considerable amount of mass concentrated in the space that it occupies. The amount of mass in the lead plus a small section of the wooden ruler on one side of your finger balances the mass existing in the longer section of the ruler on the opposing side. The same situation occurs when you balance a metal hammer with a wooden handle horizontally on your finger. The balance point would be close to the head of the hammer.

An athlete's body is similar to the hammer because it is not made of the same material throughout and the body mass is not uniformly distributed from the athlete's head to his feet. Instead, an athlete's body is made up of different substances, such as bone, muscle, fat, and tissue, all of which have different densities and different shapes. **Density** refers to the amount of substance (or mass) contained in a particular space (i.e., quantity of mass per unit volume). Bones and muscles are denser and have more mass concentrated in the space they occupy than does an equal volume of body fat.

- Therefore, the attraction between the earth and a cubic centimeter (or a cubic inch) of bone or muscle is greater than that between the earth and the same volume of body fat.

- The result of these variations in density is that an athlete's center of gravity is seldom equidistant from his head and his feet.

- But even though an athlete's center of gravity is not equidistant from head to feet, it "positions itself" so that as much mass will be directly above the center of gravity as directly below it and as much mass will be to the left of the center of gravity as to the right.

Shifts in the Center of Gravity

Living beings can shift their center of gravity from one position to another, and training teaches athletes to position their center of gravity in certain ways to produce optimal performance. During a golf drive, when athletes shift their body weight from the rear foot to the forward foot, they are in essence shifting their center of gravity so that they have optimal stability for the application of force to the ball. You will see the same repositioning of an athlete's center of gravity in all sport skills, whether they are performed in the air, on land, or in the water.

- When males stand in the anatomical position (upright with arms by the side and palms facing forward), their center of gravity is around the belly button, or navel. This point will naturally move up or down depending on the amount of mass above (such as the chest and shoulders) or below (such as the thighs).
- Because of the general difference in anatomy between males and females, the center of gravity of female athletes is usually a little lower.
- The reason for the difference is that males tend to have more body mass in the shoulders and less in the hips, whereas for females the opposite occurs.

Several factors cause the center of gravity of athletes to shift from the average positions that we've just indicated. An athlete who naturally has long and heavily muscled legs and a lighter build in the upper body will have a center of gravity that is positioned lower on the body than average. Through training, athletes can change the position of their center of gravity.

For example, a bodybuilder who works out for years on his upper body and neglects to develop his legs will shift his center of gravity upward.

- But by far the most important factor is that all of us can maneuver our center of gravity through the movement of our limbs.
- If an athlete stands erect and then moves a leg forward to take a step, the center of gravity shifts in the same direction.
- If the athlete moves the leg plus an arm, the center of gravity shifts forward even farther.

The distance that an athlete's center of gravity shifts from one position to another depends on how much of the athlete's body mass is moved and how far it's moved. Legs usually have a lot of mass, so moving them causes a greater shift in the center of gravity than moving one arm by itself. Flexing at the waist shifts the center of gravity, as does tilting the head. The shift of an athlete's center of gravity always relates to the amount of mass and the distance that it is moved (see figure 7.2).

FIGURE 7.2   An athlete's center of gravity shifts as the body's position changes.

What happens if a weightlifter hoists a heavy barbell above the head? In this situation, consider the combined center of gravity of the weightlifter's mass plus that of the barbell. Hoisting a heavy barbell to arm's length above the head shifts the combined center of gravity of athlete and barbell a considerable distance in a vertical direction (Korkmaz and Harbili 2016). In addition, the weightlifter has raised the mass of the arms above the head. The longer and more massive the athlete's arms are and the more massive the barbell is, the farther the combined center of gravity will move (see figure 7.3).

FIGURE 7.3   Combined center of gravity for a weightlifter with a heavy barbell overhead.

If the weightlifter lets go of the barbell, then the center of gravity immediately reestablishes itself relative to the body position. The barbell, of course, will have a center of gravity of its own. Is it possible to shift the center of gravity outside the body? Yes, and the more flexible the athlete is, the easier it is. A diver performing a toe touch in a pike position reaches forward with the arms and flexes at the waist. This position causes the diver's center of gravity to move forward to a position where it is no longer within the body (see figure 7.4).

A gymnast performing a high back arch or a back walkover also shifts the center of gravity into a position where it is temporarily outside the body. The center of gravity moves in relation to the shift in mass of the legs, upper body, and arms. The more extreme the arch is, the greater the shift of the center of gravity is (see figure 7.5). A gymnast's back arch position is much like the draped

FIGURE 7.4   Center of gravity for a diver in a pike position.

FIGURE 7.5   Center of gravity for a gymnast during a back walkover.

layout position that a flop high jumper uses when clearing the bar.

Just where an athlete should position the center of gravity depends on the demands of the sport skill that the athlete is performing.

- Athletes in wrestling and judo constantly reposition their center of gravity to increase their stability relative to the actions of their opponents.

- In the set position in sprint races, athletes shift their center of gravity in the direction they want to sprint so that they waste no time getting up and out of the blocks when the gun goes off.

- Gymnasts balancing on the beam make sure that their center of gravity stays centered above the beam.

- In track and field throwing events, athletes position their center of gravity so that they can apply the greatest amount of force to the implement over the greatest time.

- In all sport skills, quality performances require precise positioning of the athlete's center of gravity.

AT A GLANCE

Angular Kinetics and Center of Gravity

- Gravity can cause and initiate rotation or limit rotation, and for many activities in daily living or sport we need to overcome the force of gravity.
- If we collectively add up all the parts of the object (human or implement), we can find a balance point, or center of gravity, the point at which the weight force is equal on either side.
- Athletes can alter their center of gravity in certain ways so that they can produce optimal performance.

How Gravity Affects Flight

In events in which the athlete is in flight for a short time (e.g., high jump, long jump, gymnastics, figure skating, trampoline, and diving), the flight path of the athlete's center of gravity is set at takeoff. Some force will be applied in a vertical direction, and some in a horizontal direction. The combined effect, or **resultant**, of these two forces sets the athlete's takeoff trajectory.

In flight, the earth's gravity pulls at the athlete's center of gravity, just as it does on the surface of the earth, and this downward tug gives the athlete's flight path its familiar parabolic (i.e., curved) shape. Once in the air, the athlete cannot alter the flight path that was set at takeoff. So if a diver makes the mistake of thrusting straight up from the board (i.e., placing all of her thrust in a vertical direction), she cannot do anything to avoid coming back down on the board (Heinen et al. 2016). Likewise, a flight path that is close to horizontal means that the diver shoots out across the pool without enough height (and therefore not enough time) to perform all the twists and somersaults required in the dive. Gravity fights against the small amount of vertical thrust that the athlete put into the dive. Crazily waving the arms and legs around like kids at the local swimming pool will not change this flight path.

A high-jump athlete who is very flexible can make his center of gravity pass under the bar while his body snakes its way over the top. The benefit of this draped-bar clearance technique can be understood by considering the same athlete's attempt to clear the bar using a squat jump rather than the flop technique. If the athlete clears 1.8 m (6 ft) using the flop technique, he would need to raise his center of gravity close to 2.44 m (8 ft) to clear the same height using a squat jump (see figure 7.6).

In contrast to springboard and tower divers, trampolinists want to rise vertically above the middle of the trampoline and then drop down to the trampoline in the same way. They prefer not to "travel" along the bed of the trampoline.

APPLICATION TO SPORT

What Do Gravity and the Center of Gravity Mean in Sport?

Because gravity is all around us, we can't avoid its force, but by understanding what the force of gravity does, we can use it to our advantage in sport. As mentioned earlier, the center of gravity is the theoretical balance point of an object (human or implement). The most effective way that athletes can work with gravity is to manipulate their center of gravity. They can do this by changing body position, by raising or lowering their arms, or by shifting their mass to one foot. By making these subtle changes and therefore moving their center of gravity, high-jump athletes, for example, can in effect reduce the net distance they need to raise their body to get over the bar. In addition, the center of gravity of an implement can be manipulated. For example, by holding a baseball bat closer to the midline or choking up on the handle, athletes are manipulating its center of gravity.

Joe Robbins/Getty Images

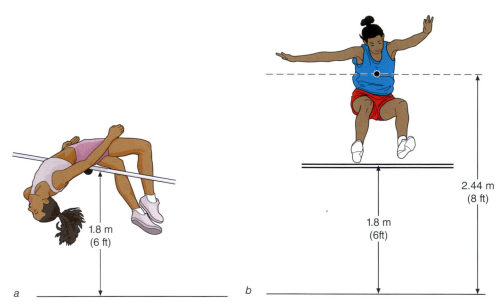

FIGURE 7.6   Height to which a high jump athlete's center of gravity must be raised to clear a 1.8 m (6 ft) bar using (*a*) the flop technique or (*b*) a squat jump.

Adapted from G.H.G. Dyson, *Mechanics of Athletics*, 8th ed. (New York: Holmes & Meier, 1986), 168.

Springboard divers train on the trampoline, but they must learn to move away from the diving board sufficiently to avoid contact. Even extraordinary divers like USA's multiple gold medalist Greg Louganis have been punished for a flight path that did not allow sufficient clearance from the end of the board. At the 1988 Olympics Louganis' head hit the diving board. He was concussed and needed stitches to his head. After being cleared by the medical staff he continued in the competition,

---

APPLICATION TO SPORT

G-Forces

The force of gravity is called g-force and is measured in Gs. Designers of roller-coasters give riders the opportunity to experience accelerations of up to 3 to 4 G in a safe way. Normal body weight is 1 G, so at 3 to 4 G, riders experience an acceleration that makes them feel three to four times heavier. There are limits to the g-forces that a healthy person can tolerate, and 3 to 4 G is considered the limit for most roller-coasters. So, exactly how many g-forces have humans experienced? In 1954 Col. John Stapp of the U.S. Air Force rode a rocket sled that reached a speed of 1,017 km/h (632 mph) in 5 s. The sled was then brought to a halt in 1.25 s. This phenomenal deceleration subjected Col. Stapp to 40 G and momentarily raised his body weight to 3,084 kg (6,800 lb)! Col. Stapp survived, and he died in 1999 at the ripe old age of 89. Since the time of Col. Stapp's exploits, humans have experienced forces of over 80 G when wearing special survival suits.

Keystone/Getty Images

Angular Momentum

Just as linear momentum describes the quantity of linear motion that occurs and is the product of mass times velocity, **angular momentum** describes the quantity of motion that a rotating athlete or an object possesses and is also the product of mass times (angular) velocity. In sport athletes often need to generate as much angular momentum as possible—to their own bodies, to an opponent, or to a bat or a club. On other occasions, athletes must reduce angular momentum to minimal values.

To help you understand how angular momentum is used in sport, let's review linear momentum and look at angular momentum. In chapter 5 we used a football lineman charging straight ahead as an example of linear momentum. The more massive the lineman is and the faster he moves, the more momentum he produces. We can apply the same concept to objects that rotate or to rotating objects (such as bats being swung) that meet other objects (such as baseballs being pitched).

In baseball a pitch travels in a predominantly linear manner to meet a bat moving in an arc. The bat being swung has angular momentum. The ball has linear momentum, and if it spins, it has some angular momentum, too (Wicke et al. 2013). Even if a pitcher hurls a fastball at 160 km/h (100 mph), the ball will still not possess much momentum because of its small mass. Because the bat has more mass, despite its swinging around its arc at a slower speed, the result is that the bat has more momentum than the ball.

You would expect this because the other way around, when the bat and ball collide, the ball would come out on top and the bat would fly out of the park. At the instant the bat and ball collide, the bat is traveling in a linear manner one way and the ball is traveling in a linear manner in the opposite direction. At impact, the momentum of the bat (bat mass × bat velocity) overwhelms the momentum of the ball (ball mass × ball velocity). The bat slows down, and the ball changes direction; it is driven backward.

Suppose a pitcher could fire the ball over the plate at 1,600 km/h (1,000 mph). Although the mass of the ball has not changed, its velocity has increased tremendously, so its momentum has increased as well. In this imaginary situation, the ball could possess more momentum at impact than the bat. The bat would be driven back, or if

it's made of wood rather than aluminum, it could snap off at the handle.

What could a coach do to compete against a pitch traveling at 1,600 km/h? Start scouting for a batter who can swing a bat at phenomenal speed! If this athlete can react fast enough to handle a 1,600 km/h pitch, the bat will win instead of the ball. In all likelihood the ball will disappear out of the park and land in the next city! In this fantasy scenario, the swing of the bat gives us some idea of the components that make up angular momentum:

- Mass (i.e., how massive the object is)
- How the mass is positioned (i.e., distributed) relative to the axis around which the object is spinning (radius of gyration)
- Rate of rotation, or swing (i.e., its angular velocity)

In reality, when bats (or clubs) are swung, there's always a tradeoff between the mass of the bat, its length and mass (weight) distribution, and the rate at which the bat is swung. In baseball, no slugger on record has used a 1.07 m (42 in.) bat, which is the limit the rules will allow. Batters also tend to choose bats weighing between 0.91 to 0.96 kg (32 and 34 oz), although no legal restrictions bar the use of heavier bats. The reason?

- A long, heavy bat with most of its mass in the barrel (i.e., hitting end) demands tremendous power from the batter and inevitably takes longer to accelerate than a light bat.
- The speed of a pitch gives a batter only a fraction of a second to react. Therefore, batters opt for lighter bats that they can swing quickly.
- Perhaps they also know that the angular velocity of the bat (i.e., its rate of swing) is more important than how massive the bat is in determining how far a baseball will travel.

Generating as much angular momentum as possible is just as important in sports in which athletes rotate as it is in sports in which they swing bats and clubs. In diving, particularly in those dives that require numerous twists and somersaults, divers generate both linear and angular momentum at takeoff. Divers use linear momentum to get high above the board as well as far enough away from the board to be safe. At the same time divers initiate rotation at takeoff and make sure that their bodies are extended and spread out. By extending and spinning, divers produce a considerable amount of angular momentum. They then use this angular momentum to perform all the somersaults and twists that occur later in the dive. Without a large amount of angular momentum initiated at takeoff, dives could not include a high number of twists and somersaults.

Increasing Angular Momentum

The batting example we looked at earlier indicates that angular momentum can be increased in three ways:

1. *Increase the mass of whatever is rotated.* Athletes can choose a heavier bat to swing. But if athletes are rotating, they must suddenly (and magically!) gain weight and spin at the same rate to increase angular momentum. Obviously, this approach is not possible.

2. *Shift as much mass as far from the axis of rotation (radius of gyration) as possible.* If athletes are rotating, they must extend their bodies. If they are swinging a bat, it must be long and have most of its mass at the barrel end.

3. *Increase the angular velocity of whatever is rotated.* Athletes can increase their **rate of spin**. Batters can swing a bat faster to give it more angular velocity.

Keep in mind that athletes always need to find a balance between mass, distribution of mass, and angular velocity. In striking and hitting skills, a huge increase in mass placed a long way from the axis of rotation is like putting a superheavy boot on the kicking foot of a field goal kicker. If the athlete doesn't possess the strength to swing his leg, the extra mass is worthless. The right amount of mass combined with a long leg and tremendous angular velocity is what is required.

In addition, an athlete, when jumping, can use angular momentum at takeoff to get higher in the air. The rules of high jump demand that athletes jump from one foot and cross a bar. A two-foot takeoff is not allowed.

- If an elite high jumper pushed down on the earth with one leg and did nothing else, the athlete would not go very high.

- A high jumper adds to the thrust of the jumping leg by initiating a strong upward swing by the arms and the free leg.

- By performing these actions, the high jumper generates angular momentum that is then transferred to the body as a whole.

- The upward swing of the arms and free leg adds to the thrust of the leg that is pushing down on the earth (see figure 7.7). More push down at the earth means that the earth in reaction pushes more against the athlete. The result is that the jumper gets higher in the air.

FIGURE 7.7 Momentum transfer in a high-jump takeoff.

In gymnastics, when gymnasts perform back somersaults on the floor, they use a technique similar to that of a high jumper. They swing their arms upward to assist in getting off the ground. For greatest effect, their arms are extended and swung upward with tremendous velocity. Elite high jumpers and gymnasts always make sure that their takeoff actions occur while they are in contact with the earth.

Figure skaters do the same. Great skaters push against the ice because the ice is part of the earth; if they push down, the earth pushes them up. The more they push, the more the earth pushes back against them. This reciprocal response from the earth is what gets skaters up in the air for their triples and quadruples. If athletes try the same actions while in the air and not in contact with the earth, a different effect occurs. We'll discuss this phenomenon in the following section.

Conservation of Angular Momentum

If an athlete or an object such as a wheel is spun, the athlete or the wheel will continue to spin forever if nothing occurs to stop the rotation. More precisely, if the turning effect of torque is applied to an athlete or wheel, the athlete or wheel will spin continuously, if not for the fact that, ultimately, torque is applied by something or someone to stop the spinning. The something or someone can be friction with a supporting surface that the athlete is spinning on, the resistance applied by air or water, or the resistance (i.e., torque) applied by another athlete.

- The principle of **conservation of angular momentum** is the rotary equivalent of Newton's first law of inertia, which tells us that mass on the move wants to continue moving in a straight line and will do so in the absence of oppositional forces.

- In a rotary sense, spinning objects will continue spinning as long nothing occurs to stop them.

In sport many instances occur when the resistance opposing the spin of an athlete is so small as to be negligible. Think of gymnasts rotating in the air as they dismount from a high bar routine. Consider high jumpers and long jumpers in flight, or divers and trampolinists rotating in flight. In all these cases, air resistance is so small that it can be discounted. In these situations we say that their angular momentum is conserved.

The word *conserve* means to stay the same, or to be maintained. This meaning applies to the amount of angular momentum that an athlete possesses during a particular phase of a skill. For example, the angular momentum that an athlete generates at takeoff in high jump, long jump, and diving remains the same while the athlete is in flight. Why?

The explanation is that an athlete cannot push against the air to increase or decrease angular momentum; during flight the air has a negligible effect in reducing angular momentum. The only force that is working on the athlete is the earth's gravitational force, which pulls at the athlete's center of gravity. This force doesn't have any effect on the athlete's angular momentum, although gravity certainly increases her linear momentum as she accelerates toward earth.

Consequently, the athlete's angular momentum doesn't increase or decrease. It stays as is. As with linear momentum, the angular momentum generated at takeoff is conserved during flight. Let's look at diving again to see how the conservation of angular momentum is tied in with a diver's ability to control the rate of spin while in the air.

Controlling the Rate of Spin in Diving

When divers accelerate down from the 10 m (33 ft) tower, they hit the water in less than 2 s. In flights of such short duration, the diver's angular momentum is conserved. Discounting the negligible amount of angular momentum lost pushing the air around, we can say that the amount of angular momentum generated by the diver at takeoff stays virtually the same throughout the dive.

In flight, divers, just like gymnasts, shift from layout body positions to tight tucks. To get to the latter position, divers use muscular force to pull the legs and arms inward, tuck in the chin, and flex the spine. By doing this, divers pull their mass closer to the axis of rotation, reducing the radius of gyration.

When this happens, rotary inertia (i.e., the diver's resistance against rotation) is reduced and the diver spins faster (i.e., the diver's angular velocity increases). But what causes this faster spin? Where does the extra angular velocity come from? We find the answer by examining what happens when one of the items that make up angular momentum is increased or decreased.

We know that the amount of angular momentum that divers possess in flight is determined by three items:

1. their rate of spin,
2. their mass, and
3. the distribution of their mass through the radius of gyration.

In other words, angular momentum is determined by how much spin they have, how much mass they have, and how extended or tucked up they are (Walker et al. 2016). All three of these factors combine to make up the divers' angular momentum.

If divers are in a situation (e.g., in flight) in which their total angular momentum stays at a set amount (i.e., is conserved) and one factor involved in creating angular momentum is reduced (e.g., they tuck and pull their body mass inward), then another component of angular momentum must increase to keep their total angular momentum unchanged. Because divers cannot possibly change their mass (i.e., gain or lose weight) while in flight, when they tuck and pull themselves in toward their axis of rotation, their angular velocity must increase. Pull your body inward, and you spin faster. Spread your body out, and you spin slower.

Controlling the Rate of Spin in Figure Skating

We see another example of athletes reducing rotary inertia when figure skaters pull themselves in tight to increase the rate of rotation when performing a multiple twisting skill in the air, such as a quadruple toe loop or a triple Axel. In these skills skaters complete several rotations (i.e., twists) around their long axis (i.e., from head to feet) during their flight. During the short time that they are in flight, their angular momentum is conserved.

A skater drives up into an Axel with one leg forward and the other back. The arms are extended sideways. As a result, the skater's body mass is spread out relative to the long axis of the body (similar to divers being extended at takeoff from the board).

- In a spread-out position, the skater's rotary inertia (resistance) is considerable.
- Consequently, the angular velocity (i.e., rate of spin) around the long axis is minimal.
- But in flight the skater pulls his arms and legs inward, which greatly reduces his radius of gyration and subsequent resistance against rotation. He now spins with tremendous angular velocity.
- When he spreads his arms and legs out again on landing, rotary inertia is again increased and the rate of spin reduced.

The principles just discussed apply even when a skater performs high-speed spins while in contact with the ice. The rate of spin increases as the limbs are pulled inward. The difference between this skill and a skill performed in the air is that a skater's blades pressing and turning on the ice generate a little more resistance than the air does when the skater is in flight. Therefore, a skater loses some

AT A GLANCE

Angular Momentum in Sport

- Angular momentum describes the quantity of motion that a rotating athlete or object possesses.
- In sport, an athlete can increase angular momentum by increasing the mass of whatever is rotated, shift as much mass as far from the axis of rotation (radius of gyration) as possible, or increase the angular velocity of whatever is rotated.
- The principle of conservation of angular momentum is the rotary equivalent of Newton's first law of inertia, which tells us that mass on the move wants to continue moving in a straight line and will do so in the absence of oppositional lifts

angular momentum because the skates experience friction with the ice.

Making Use of Angular Momentum

When tower or springboard divers are in flight, they no longer have a large mass to push against. Any muscular action they perform while in the air causes an equal and opposite reaction to occur elsewhere in their bodies. All divers, gymnasts, and other high-flying athletes experience this phenomenon.

For example, imagine a diver stepping off the tower and dropping toward the water in an upright position. In this position, he is not rotating. As the diver drops, imagine the coach shouting for him to raise his extended legs 90° from perpendicular (i.e., pointing directly downward) to horizontal. The muscles that rotate his legs forward and upward around the hip joint pull equally at both origin and insertion (i.e., at either end of the muscle) and therefore simultaneously pull down on his trunk. As a result, during the period that the diver's legs rotate upward, his trunk must rotate downward. Do his legs and trunk rotate in equal-size arcs?

No, because they do not have the same rotary inertia. The rotary inertia of the diver's trunk and upper body is approximately three times that of his legs. Therefore, his trunk and upper body resist rotation three times more than his legs do. When the diver's legs move upward 90° to a horizontal position, the trunk and upper body, which have three times more rotary resistance, move downward in an arc a third of the size, or approximately 30° (see figure 7.8).

The movement of the diver's legs and that of his trunk and upper body may not seem like equal and opposite reactions, yet they are. In our example, the action is the 90° arc moved in a counterclockwise direction by the diver's legs. The reaction is the 30° arc moved in a clockwise direction by their trunk and upper body. This reaction is equal to the action because the diver's trunk and upper body have three times the rotary inertia of the legs. The reaction is opposite because the movement of the trunk and upper body is in the opposing direction to that of the legs.

We've seen that rotary inertia depends not only on how much mass is involved but also on how it is distributed relative to its axis. In the previous example, the diver keeps his legs extended throughout their 90° movement. If the diver flexes at the knees and lifts at the thighs, then the rotary inertia of the legs is reduced. The reaction of the trunk and upper body is also reduced. The upper body and trunk flex approximately 20° (see figure 7.9).

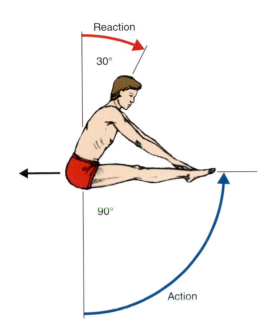

FIGURE 7.8 When the diver's extended legs are raised 90° in a counterclockwise direction, the upper body reacts by moving 30° in a clockwise direction.

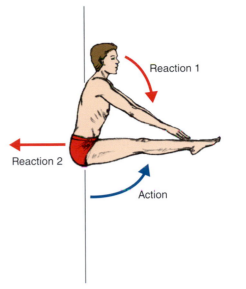

FIGURE 7.10 In the air, when the lower and upper body flex forward, the hips react by shifting backward

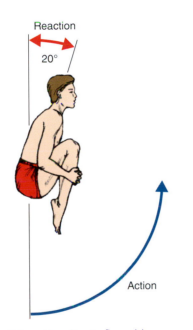

FIGURE 7.9 When the diver's flexed legs are raised, they have less rotary inertia than when they are raised in an extended position. The upper-body response is less.

Figure 7.10 shows that another reaction will occur in the diver's movements. Notice that as the diver's extended legs rotate counterclockwise, the trunk and upper body rotate clockwise. In the illustration, these body parts move toward the right as you look at them. In the air, body mass moving to the right is counterbalanced by body mass moving to the left. In our example the diver's buttocks and hips react (equal and opposite) by moving toward the left.

This phenomenon of equal and opposite reactions occurring in the air is visible in many sports. A flop-style high jumper always arches to clear the bar. Figure 7.11 shows a high jumper arching the upper body down toward the pit in a counterclockwise direction. The legs respond (equal and opposite) by moving in a clockwise direction. Although upper body and legs are moving in opposing directions, both are moving down toward the pit. So the hips react by moving upward. By correctly timing this action, the high jumper can pass over the bar. The hips and buttocks move upward to help the athlete clear the bar.

Another example of this phenomenon occurs when a volleyball player jumps to spike a ball. As the upper body flexes backward in a counterclockwise direction, the lower body reacts by moving in a clockwise direction. As shown in figure 7.12, even though the upper body and lower body are rotating in opposing directions, both move to the left. An equal and opposite reaction must balance this movement. The hips and abdomen shift in the opposing direction. Do you see how this replicates the action of the high jumper?

In events such as the long jump, high jump, figure skating, diving, and gymnastics, the athletes are in the air for a short time. During flights of short duration their total angular momentum

FIGURE 7.11   The high jumper's hips lift upward when the upper and lower body flex downward.
Ty Allison/Photographer's Choice/Getty Images

FIGURE 7.12   Action and reaction in a volleyball spike.

remains constant. If athletes introduce more angular momentum during flight by suddenly rotating the arms or legs, another part of the body (or the body as a whole) must give up some angular momentum to keep the total constant.

In other words, if a long jumper gives herself 10 units of angular momentum at takeoff, she cannot increase the 10 units to 12 while in flight. If the long jumper introduces 2 units of angular momentum by rotating her arms and legs, then 2 units must disappear elsewhere in her body to keep the total constant at 10. Interestingly enough, the principle of introducing angular momentum with the arms or legs to get rid of angular momentum elsewhere in the body gives athletes some control over their movements while in the air. Let's first look at what occurs in ski jumping; then we'll take a closer look at the long jump, where elite jumpers put this principle to good use.

Controlling Forward Rotation in Ski Jumping

In ski jumping, a jumper who mistakenly gives himself too much forward rotation at takeoff knows that unless he does something about the problem, he is likely to land on his face! So the ski jumper rotates his arms in the same direction as the unwanted forward rotation. The angular momentum generated by the arms helps put the brakes on the forward rotation of his body. If performed vigorously enough, the arm action can rotate the skier's body backward into a more favorable position. If the skier rotates his arms backward, his body will rotate farther forward, which in the situation we've described is the last thing he wants to do.

Controlling Forward Rotation in the Long Jump

All long jumpers rotate forward at takeoff. Even world-class jumpers face this problem. Forward rotation is impossible to avoid because the takeoff foot pushing back at the board causes the athlete's body to rotate in the same direction as the direction of push of the takeoff foot. A long jumper taking off from left to right pushes toward the left with the takeoff foot. In this case, the jumper will have given the body clockwise rotation. Worse still, the body continues to rotate forward throughout flight unless the athlete does something about it. The legs and feet will rotate down toward the sand, and the distance jumped will be greatly reduced. Novice jumpers who hold a bunched-up position in flight get themselves in trouble because a tightly flexed position means that they also spin quickly. Thus, a novice's body and feet quickly rotate down toward the sand, resulting in a mediocre distance (see figure 7.13).

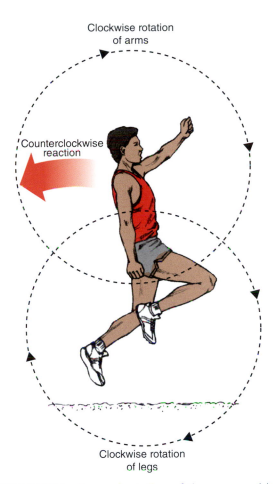

FIGURE 7.14 Forward rotation of the arms and legs causes backward rotation of the long jumper's body.

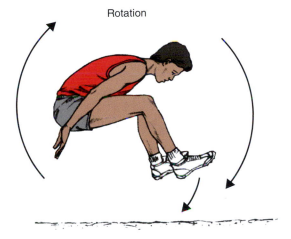

FIGURE 7.13 In the air, a bunched position increases the long jumper's angular velocity, so the feet hit the sand earlier than necessary.

To counteract unwanted forward rotation initiated at takeoff, an expert long jumper rotates the arms and legs in the same direction (i.e., forward) while in the air. Elite jumpers who have perfected the cycling action of a "hitch kick" can stop their bodies from rotating forward and make them rotate in the opposing direction (see figure 7.14). This change in rotation helps them achieve a good body position for landing and a greater distance jumped.

Just how much reaction an athlete gets from rotating the arms and legs depends on how much angular momentum the athlete's body has been given at takeoff and how much angular momentum is introduced by the rotary motion of the arms and legs. Vigorous rotary actions with arms and legs extended produce the most angular momentum. Look for expert long jumpers to flex their legs to drive off the ground and to rotate their arms forward to increase their momentum (Allen et al. 2016). This action has maximum effect on counteracting the forward rotation of the athletes' bodies.

A good place to see desperate use of angular momentum is at your local swimming pool. Youngsters run and jump off the springboard. They don't expect the board to flip their feet upward to the rear, and in horror they find themselves rotating in the air and heading for a belly flop. It is at this point that they introduce a wild flailing of the arms. These youngsters, uneducated in applied sport mechanics, are introducing angular momentum in the correct direction. They rotate their arms, and frequently their legs, in the same direction as the unwanted rotation they received from the springboard. Without knowing it, they replicate in rough form the precise movements of an elite long jumper.

Transferring Angular Momentum

Divers, gymnasts, and ski aerialists frequently combine somersaults with twists. In these complex skills, athletes simultaneously somersault around their transverse axis (hip to hip) and twist around their long axis (feet to head). The most remarkable technique used to combine somersaults with twists is one in which athletes begin by somersaulting with no twist apparent. Let's look at diving to examine the mechanics of these actions.

- At takeoff, divers push against the board to help get their bodies rotating and thus generate angular momentum. As with long jumpers, the amount of angular momentum produced at takeoff remains virtually the same throughout the diver's flight down to the water.
- In flight, the diver begins by rotating around the somersaulting axis (i.e., transverse, or hip to hip).
- Then the diver performs a series of body actions that borrow some angular momentum from the somersault to put into the twist.
- The diver now somersaults and twists.
- After somersaulting and twisting for the required number of revolutions, the diver performs a second series of actions.
- These actions remove the twist, and the diver somersaults without twisting.
- The dominant method for borrowing, or stealing, angular momentum from the somersault and placing it in the twist is called the body tilt technique.

Body Tilt Technique

The body tilt technique of twisting requires the diver to tilt the body away from its somersault axis so that some angular momentum goes into the twist axis. This technique is best understood by imagining the diver as a solid block somersaulting around the transverse (somersault) axis. The diver's angular momentum while in flight is set at takeoff, as shown in figure 7.15.

- In figure 7.15a, a specific amount of angular momentum is given to the block around its transverse axis.

- Suppose that during the somersaulting rotation, the block is tipped sideways so that it lies horizontally, as shown in figure 7.15b. Rotation continues, but the block now rotates around its longitudinal axis. In effect, the block's long axis has become its somersault axis.
- Now let's tilt the block over just a few degrees, as shown in figure 7.15c, rather than all the way to horizontal. The block will twist and somersault at the same time. Why?

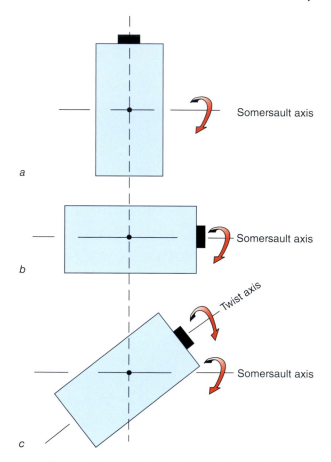

FIGURE 7.15   Mechanics of the body tilt technique of twisting.

The angle of tilt forces some angular momentum into the twist axis (i.e., the block's long axis), but some still remains in the original somersault axis. In addition, the rotary inertia (i.e., resistance) of the block around its twist axis is less than that around the somersault axis. The block twists faster than it somersaults. If the block is brought back to its original position, the process is reversed. The twist disappears and all angular momentum goes back into the somersaulting axis.

The principles of the body tilt technique indicate that if divers start a dive by somersaulting, they will somersault and twist if they tilt over slightly while in flight. But how do divers tilt themselves over while in the air? Here's how it occurs.

- When divers take off from the board, they extend their arms sideways (see figure 7.16*a*).
- Then, when in flight, they swing one arm downward and the other simultaneously upward (see figure 7.16*b*).

Notice that even though one arm goes up and the other down, both arms rotate clockwise. The angular momentum generated by these clockwise arm actions causes the diver's body to react by rotating in a counterclockwise direction. A double arm swing of 90° (one arm swung up 90° and the other swung down 90°) causes the body to tilt approximately 5° from vertical, sufficient for a diver to initiate a twist. Swinging the arms back reestablishes the original body position and eliminates the twist.

A common sequence for the combination of a somersault and twist using body tilt is as follows. The athlete begins the dive by initiating a somersault at takeoff. In flight, the diver raises one arm and lowers the other. The diver's body tilts over, which causes twists and somersaults to occur at the same time. When the diver returns the arms to their original positions, the twist is eliminated. The somersault action continues and is reduced to a minimum when the diver's body is fully extended for entry into the water.

Divers are also aware that the more body tilt they can establish, the more twists they can perform. So what can they do to exaggerate the body tilt? After a half twist around the long axis is complete, the arm that was lowered is raised and wrapped to the rear of the head. The other arm, which had been raised, is lowered and pulled in tight across the waist. This repetitive arm action tilts the diver over farther, causing twists to occur at an incredible rate and making dives with triple twisting somersaults possible.

Cat Twist Technique

This twisting technique is called the cat twist because a cat, like many other animals, performs the mechanics of this technique naturally, without training or coaching. Experimentation has shown

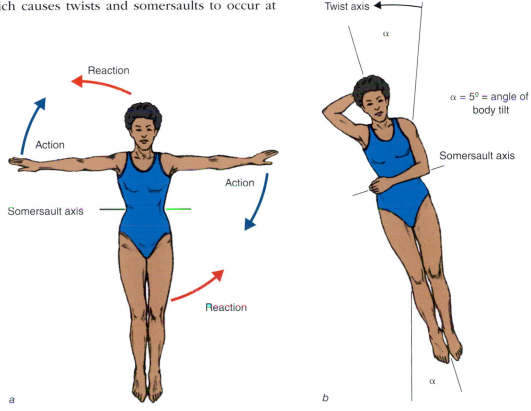

a

b

FIGURE 7.16   (*a*) A diver with arms outstretched rotates around the somersault axis. (*b*) When the diver raises one arm and lowers the other, her body tilts. This causes some somersault momentum to be transferred to the twist axis.

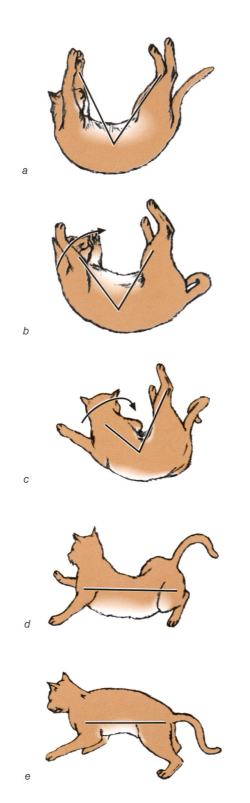

FIGURE 7.17 Cat twist. The cat reduces the rotary inertia of the upper body and twists the upper body against the larger rotary inertia of the lower body. Notice the angle of the body in *a* through *c*.

that if a cat is held inverted in a static position 1 m (3 ft) from the ground and allowed to drop, the animal can initiate a twist in the air so that it can land safely on all fours. Figure 7.17*a* through 7.17*e* shows a cat as it falls. Notice the piked angle at the cat's midsection in figure 7.17*a*.

This important angle allows the cat to twist the upper body against the rotary inertia of the lower body and vice versa. The first move that the cat makes is to establish visual contact with the ground.

- The cat pulls in the forelegs and turns the upper body toward the ground (see figure 7.17*b* and 7.17*c*).
- By pulling in the forelegs, the cat reduces the rotary inertia of the upper body in relation to its lower body, which has more rotary inertia because the rear legs are fully extended.
- Consequently, the upper body, with its smaller rotary inertia, can be rotated through a large angle with minimal reaction in the opposing direction from the lower body.
- When the cat sees the ground, it extends the forelegs in preparation for landing (see figure 7.17*d*). This action increases the rotary inertia of the upper body.
- The cat pulls the rear legs inward, reducing the rotary inertia of the lower body and allowing the lower body to rotate against the greater rotary inertia of the upper body. Once fully aligned, the cat is ready to land safely (see figure 7.17*e*).

The cat twist differs from the body tilt technique because it can be performed without borrowing, or stealing, angular momentum from a somersault. In other words, an athlete does not have to be somersaulting to perform the cat twist action. The cat twist always requires the athlete's body to have an angle, or to be bent at the waist in some way. The direction doesn't matter. Flexion at the waist can be forward, backward, or sideways—they all work successfully. A 90° angle at the waist is preferable (i.e., with the upper body piked forward), but the cat twist can be performed with far less than 90°. An elementary skill on the trampoline that employs the cat twist technique is the swivel hips.

The piked position used in the swivel hips also occurs in diving. Figure 7.18*a* shows a diver in a piked position with a right angle at the hips. In this position, the mass of the legs is a long way out relative to the long axis, which extends from the athlete's hips to the head. Figure 7.18*b* shows

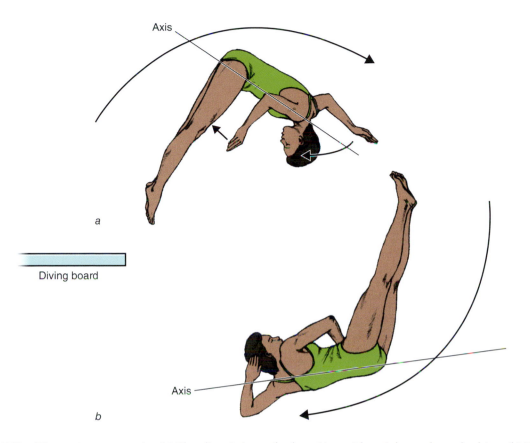

FIGURE 7.18   Diver using a cat twist. (*a*) The diver is in a piked position with a right angle at the hips. (*b*) The diver twists the upper body against the greater rotary inertia of the legs.

Adapted from M. Adrian and J. Cooper, *The Biomechanics of Human Movement* (Indianapolis, IN: Benchmark Press, 1989), 694, by permission of The McGraw-Hill Companies, Inc.

the athlete twisting the upper body against the greater rotary inertia of the legs. This action is the basic principle of the cat twist. A twist is achieved by turning one part of the body that has reduced rotary inertia against another part that has much larger rotary inertia.

The cat twist is frequently used at the start of forward dives that contain somersaults and twists. In these dives the athlete combines aspects of both the cat twist and the body tilt technique. Figure 7.19*a* through 7.19*e* shows a diver taking off for a dive that will contain both somersaults and twists. Notice in figure 7.19*a* that the diver's arms are extended outward above the shoulders. Rotation around the transverse somersault axis is initiated at takeoff. Figure 7.19*b* shows the 90° pike required by the cat twist and the raising of one arm and the lowering of the other using the principles of the body tilt. After the upper body is twisted against the inertia of the lower body, the angle at the waist is removed (see figure 7.19*d* and 7.19*e*) to allow

the twist to continue around the long axis of the diver's body.

AT A GLANCE

Transferring Angular Momentum in Sport

- Based on the fundamental law of equal and opposite reactions, an athlete can transfer angular momentum from one part of the body to another.

- The key components to transfer angular momentum are mass and angular velocity, in particular how much mass is involved and how this mass is distributed relative to its axis of rotation.

- The transfer of angular momentum can travel across different planes of moment (e.g., transverse, longitudinal).

FIGURE 7.19 A diver simultaneously using the cat twist and body tilt technique after takeoff.

Divers and ski aerialists use other twisting and somersaulting techniques. Unfortunately, a discussion of these techniques is beyond the scope of this text. If you coach or intend to coach gymnastics, diving, figure skating, ski aerials, trampoline, or any sport in which athletes twist and turn in the air, be sure you understand the mechanics of the event and always use proven safety techniques.

SUMMARY

- Angular momentum is the rotary equivalent of linear momentum: It describes the quantity of rotary momentum. The angular momentum of an object is determined by the product of its mass, its angular velocity, and the distribution of its mass.
- In flights of short duration (e.g., diving, long jump, gymnastics, and high jump), the amount of angular momentum generated by an athlete at takeoff remains the same for the duration of the flight. The athlete's angular momentum is conserved.
- When an athlete's angular momentum is conserved, the athlete's rate of spin (angular velocity) increases or decreases in relation to changes in the distribution of the athlete's mass. For example, the tighter the tuck is, the greater the angular velocity is.
- In flight, the angular velocity of an athlete's body increases or decreases in relation to the introduction of angular momentum by other body parts. During flight, clockwise rotation of the athlete's body as a whole can be counteracted by the introduction of clockwise rotation in the athlete's arms and legs. The same principle applies to counterclockwise rotation.
- Divers use combinations of the body tilt, cat twist, and other techniques to twist and somersault.
- The body tilt technique of twisting requires athletes to be somersaulting before initiating the twist. This technique uses specific arm actions to tilt the body out of the somersault axis. The action transfers angular momentum from the somersault axis to the twist axis. Athletes then twist and somersault simultaneously. When the body tilt is removed, the twist is eliminated and the somersault continues.
- The cat twist technique can be initiated in flight with zero angular momentum. No somersault needs to be initiated beforehand. This technique requires athletes to be flexed at the waist. The athlete twists by rotating one part of the body that has reduced rotary inertia against another part that has larger rotary inertia.

KEY TERMS

angular kinetics

angular momentum

balance point

center of gravity

conservation of angular momentum

conserve

density

gravity

gravitational acceleration

rate of spin

resultant

transfer of angular momentum

REFERENCES

Allen, S.J., M.R. Yeadon, and M.A. King. 2016. "The Effect of Increasing Strength and Approach Velocity on Triple Jump Performance." *Journal of Biomechanics* 49 (16): 3796-802.

Haase, S.C. 2017. "Management of Upper Extremity Injury in Divers." *Hand Clinics* 33 (1): 73-80.

Heinen, T., M. Supej, and I. Čuk. 2016. "Performing a Forward Dive With 5.5 Somersaults in Platform Diving: Simulation of Different Technique Variations." *Scandinavian Journal of Medicine and Science in Sports* 27 (10): 1081-89.

Korkmaz, S., and E. Harbili. 2016. "Biomechanical Analysis of the Snatch Technique in Junior Elite Female Weightlifters." *Journal of Sports Sciences* 34 (11): 1088-93.

Walker, C., P. Sinclair, K. Graham, and S. Cobley. 2016. "The Validation and Application of Inertial Measurement Units to Springboard Diving." *Sports Biomechanics*: 1-16.

Wicke, J., D.W. Keeley, and G.D. Oliver. 2013. "Comparison of Pitching Kinematics Between Youth and Adult Baseball Pitchers: A Meta-Analytic Approach." *Sports Biomechanics* 12 (4): 315-23.

Visit the web resource for review questions and practical activities for the chapter.

8

Stability and Instability

Minas Panagiotakis/Getty Images

When you finish reading this chapter, you should be able to explain

- the importance of stability in sport skills;
- how athletes make use of linear and rotary stability;
- the mechanical principles that determine different levels of linear and rotary stability;
- why some sport skills require minimal stability;
- why the rotary stability of a spinning object is proportional to its rotary resistance; and
- how to measure center of gravity and the line of gravity.

This chapter discusses the importance of balance and stability in the performance of sport skills. You'll see why some sport skills require athletes to maximize their stability, whereas other skills require athletes to reduce their stability to minimum levels temporarily. You'll also see how stability is related to mass and inertia and, in particular, how preservation of an athlete's stability is always a battle of torques. The turning effect of one torque that disrupts an athlete's stability is battled by the turning effect of torque applied by the athlete's muscles to regain stability. Stability and instability are powerful sport techniques that you can use when applying your knowledge of science to sport. To help you put this into practice, this chapter first establishes the mechanical principles for stability and the factors that determine stability. By altering the base of support or the **line of gravity**, we can change the stability or instability of the athlete.

Many athletes naturally sense how to move and seem to know instinctively what they should do to maximize their stability. Unfortunately, not all athletes are gifted in this way. Young athletes, particularly those struggling to learn a new skill, assume poor body positions, which reduce the quality of their performances.

- They don't plant their feet properly when throwing, striking, or hitting, and they find that the reaction forces resulting from their actions cause them to stumble or "fly" in the opposite direction.
- These athletes don't assume efficient stances when checked, blocked, or challenged by an opponent.
- When they become unstable, or lose their balance, they don't make the correct maneuvers to regain control quickly.
- If they want to move suddenly, they are unable to do so because their stance simply doesn't allow them to get off the mark quickly.

For ease of reading, the terms *athlete* and *object* can be interchanged; that is, if we are talking about an athlete's stability, the same principles apply to an object. All these examples are errors involving the amount of stability required at a particular instance in the performance of a sport skill. Most of these errors are easy to correct provided you understand the mechanical principles relating to stability. You'll read about these principles in this chapter.

Fundamental Principles for Stability

Most activities of daily living require human beings to control their stability; that is, they can maintain their current static position, or they can move in the direction they desire (Punt et al. 2015). The mechanical definition of **stability** specifically relates to how much resistance the human can put up. The more stable a human is, the more resistance he can generate against any disruptive forces.

When a world champion gymnast works flawlessly through a routine on the beam, she successfully counteracts the forces that would cause her to fall off. Compare the gymnast with National Football League running backs. These athletes twist and turn and keep driving for the end zone even when they are bombarded with repeated tackles from their opponents. In spite of the great differences in their sports, world-class gymnasts and elite running backs both demonstrate superior control of their stability.

We can use the broad term **balance** to refer to the ability to control the forces that make an athlete stable. This term simply means that the athlete has control of her stability and can manipulate her stability in a controlled manner. An athlete may require stability in a static or a dynamic situation.

For example, to perform a handstand, athletes control the stability forces to minimize any movement as they remain as still as possible. In a highly dynamic scenario, such as when a running back drives for a touchdown, players want to maintain stability as they encounter the dynamic tackling forces of the opposition, along with the usual adversaries of gravity, friction, and air resistance.

Mechanical Principles for Stability

From a mechanical perspective we can define stability using simple geometric shapes. For example, a triangle with one edge on the ground is stable (see figure 8.1*a*); alternatively, a triangle sitting on an apex is unstable (see figure 8.1*b*); and finally, a circle is neutral (see figure 8.1*c*). The influence of stability on applied sport mechanics can be subdivided into two broad categories:

- **Linear stability** is concerned with an athlete who is moving in a particular direction.
- **Rotary stability** is concerned with an athlete who is rotating.

Linear Stability

Athletes (or objects) can put up a certain amount of resistance against being moved in a particular

direction. The same athletes can possess a certain amount of resistance against being stopped or having their direction changed once on the move. Both situations are types of linear stability. For example, the sumo wrestler can resist being forced by his opponent in a particular direction. Then when he charges across the ring to slam into his opponent, his opponent needs to apply force to stop or change his direction.

If a 136 kg (300 lb) sumo wrestler bulks up to a massive 181 kg (400 lb), then he will need more strength to overcome his own inertia and get himself moving. At the same time, an opponent will require more force to move the more massive sumo wrestler from a static and stable position or to change his direction when he is charging across the ring. Consequently, the more massive the sumo wrestler is, the more inertia he has and the more stability he has too. So linear stability is directly related to mass. Sumo wrestlers are well aware of this fact. Therefore, they lift weights to become more powerful and stuff themselves with food to become more massive.

Whether athletes want to have great linear stability or not depends on the demands of the skill. In a rowing race, athletes want to row the shortest

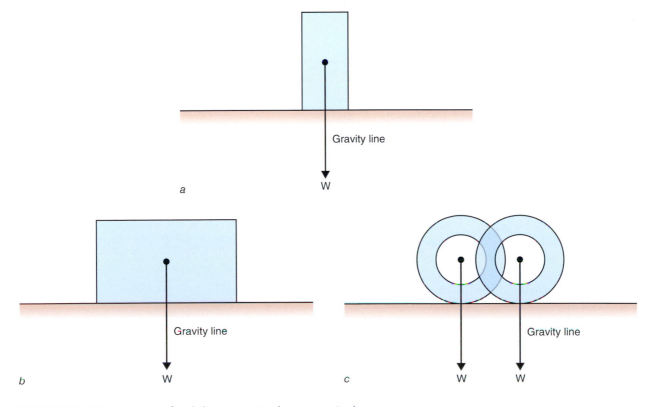

FIGURE 8.1   The concept of stability using simple geometric shapes.

Adapted from E. Kreighbaum and K. Barthels, *Biomechanics: A Qualitative Approach for Studying Human Movement*, 4th ed. (Upper Saddle River, NJ: Pearson Education, 1996), 332.

distance between the start and the finish. To do this, they try to counteract any forces that might shift them off course. A heavyweight rowing eight powering toward the finish has tremendous linear stability. Collectively, the eight rowers and their rowing shell form a huge mass. The long narrow shape of their shell, coupled with their rowing actions, propels them at high speed in a straight line. The adversary trying to push them off their straight-line course or slow them down is predominantly wind, waves, and friction generated by the shell and the rowers moving through the water (and to a lesser extent through the air).

At the other end of the spectrum, surfers, figure skaters, and slalom skiers differ from a rowing eight because these athletes must contend with sudden directional changes. They also want a certain amount of linear stability, but they don't want to keep going in a straight line when it's necessary to make sudden tight turns. Therefore, athletes who want to shift and turn quickly don't want the weight of a sumo wrestler. Imagine a slalom skier trying to maneuver 181 kg (400 lb) of body weight though a series of tight turns (Falda-Buscaiot et al. 2017)! It's no different for squash players, badminton players, and goalkeepers in soccer. Tremendous body mass means great inertia, and great inertia needs a huge amount of physical power to initiate movement. Once on the move, great inertia implies straight-line movement. Therefore, too much mass means too much stability, which can be a liability in skills in which high-speed movements, first in one direction and then the other, are required.

In addition, friction plays a role in sport whether athletes are stationary or are moving. To help increase stability when athletes are stationary, frictional forces are important. Consider a massive wrestler in the Olympic heavyweight division lying in a defensive position on the wrestling mat. Lying flat and spread-eagle, the athlete is difficult to move. The wrestler's huge mass is pulled down tight onto the mat by gravity. You can see why an opponent would have to be phenomenally powerful to pull, push, or roll this athlete out of his defensive position.

Compare this scenario with what is required of a speed skater. Speed skaters are strong, muscular athletes, but they are considerably lighter than heavyweight Olympic wrestlers, and the frictional differences between a wrestling mat and slick ice are obvious. The slightest backward thrust from the skater causes the skate blades to glide over the ice with minimal frictional resistive force. The static stability of the speed skater in this situation is much less than that of the wrestler.

Regardless of the surface type, a heavy, massive athlete naturally presses down onto a supporting surface more than a lighter athlete does. In a sport like football, the pressure generated by the mass of a huge lineman in contact with the earth produces more friction between his cleats and the turf than the pressure generated by a lighter running back. The lineman therefore has better traction than a running back (of course, the lineman has considerably more mass and inertia than a running back, so although he may have more traction he certainly doesn't have a running back's maneuverability). An opponent must overcome the lineman's friction with the turf if he wants to block him out of a play.

Rotary Stability

Rotary stability is the resistance of an athlete (or an object) against being tilted, tipped over, upended, or spun around in a circle. But if the turning effect of a torque is sufficient to set an athlete (or an object such as a discus) spinning, rotary stability then describes the ability of the athlete and the object to keep spinning and resist whatever would slow down the rate of spin. We'll look at the principles that help an athlete avoid falling or being tipped over, upended, or spun around. You'll notice that these important principles are universal in sport.

The amount of effort required to maintain balance varies from sport to sport. An elite sprinter is normally not thinking about stability during a 100 m race, and a top-class basketball player doesn't expect to fall over when bringing the ball up court. For other sport skills, however, athletes must actively work to maintain stability. For example, in judo, athletes battle to maintain their stability while trying to destabilize their opponents as they attack and counterattack (Barbado et al. 2016). In weightlifting, athletes struggle to control immense barbells held at arms' length above their heads, and in sprint cycling, you'll often see competitors balancing almost motionless as they try to outmaneuver each other at the start of a sprint race in the velodrome. All these athletes are maintaining various levels of rotary stability.

- Excellent rotary stability makes athletes better able to resist the destabilizing and turning effect of a torque applied against them.
- The more stable they are, the more torque is necessary to upset their balance.

Let's see how this battle of torques plays out. A destabilizing torque that upsets an athlete's stability can come from any external source. It can be generated by gravity, air resistance, an opponent, or any combination of forces. The axis around which athletes rotate when a torque is applied against them can be any place on their body, an opponent's body, or some external object.

- If a gymnast loses her balance on the beam, the axis of rotation will be where her body contacts the beam. It can be the ball of one foot during a pirouette or her hands during a handstand.

- In a hip throw in judo, the point of rotation is where the athlete's body contacts the opponent's hip.

- If a sprinter stumbles during a lunge for the tape, the point of rotation is likely to be where his spikes contact the track.

When a gymnast performs a difficult one-handed handstand, she is balanced but not very stable. She needs to fight to maintain her balance. If she allows her body to shift even the slightest distance out of a balanced position, a turning effect (or torque) occurs as she starts to rotate. The earth pulls at her center of gravity. The axis of rotation will be where her hand contacts the floor or the apparatus. The farther she shifts out of a balanced position, the greater the turning effect will be. In a one-handed handstand, a gymnast must use the strength in her supporting hand and forearm (plus other muscles in her body) to counter a shift out of a balanced position. She does this by producing an opposing torque. If the turning effect of this opposing torque is powerful enough, it will rotate the gymnast in the opposite direction until a state of balance is regained. But if she allows her center of gravity to shift too far from its correct position (which is directly above the supporting hand), then she is not likely to have the strength to pull herself back to a balanced position. Gravity then wins this battle of torques, and the gymnast collapses out of the handstand. Figure 8.2 shows a gymnast who has allowed her center of gravity to shift too far to the left. The greater this distance (d) is, the greater the torque produced by gravity is.

Compare a gymnast balancing in a one-handed handstand with either of the two Greco-Roman wrestlers shown in figure 8.3. Obviously, neither of the two wrestlers has any stability at all. The defender is being thrown and is totally out of contact with the mat. The attacker has only one foot solidly in contact with the mat, and his center of gravity and line of gravity are both well outside his supporting base. In this highly unstable position,

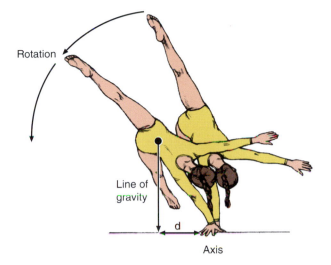

FIGURE 8.2 When the center of gravity is no longer above the supporting base, gravity applies a destabilizing torque. The greater the distance (d) the line of gravity is from the axis, the greater the torque produced by gravity is.

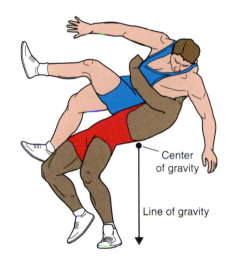

FIGURE 8.3 The attacking wrestler in contact with the mat has his center of gravity well outside his base. In this unstable situation, he has put himself at risk to throw his opponent.

he has gambled everything to throw his opponent, and his gamble has paid off!

Figure 8.4, another example from wrestling, shows how rotary stability is a war of one torque versus another. The torque applied by the attacker (i.e., force × force arm) competes against the torque (i.e., resistance × resistance arm) generated in the opposing direction by the defending wrestler. In this situation the defending wrestler's body weight acts as the resistance, and his hand is the axis of rotation. The attacker tries to increase the turning effect he is applying to his opponent by increasing both the force he is applying and the length of the force arm he is using. The defending wrestler obviously cannot increase his rotary stability by gaining body weight during the wrestling match. But what he tries to do is to keep his resistance arm as long as possible. In this figure, the defending wrestler could lengthen his resistance arm even more by shifting his hand on the mat farther away from his opponent (remember that the word *arm* in this context refers to the perpendicular distance from his center of mass to the axis of rotation, not to the length of his actual arm).

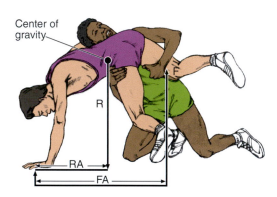

FIGURE 8.4   Wrestling is a battle of torques. The attacker's force (F) × force arm (FA) competes against the defender's resistance (R) × resistance arm (RA).

Let's stay with the sport of wrestling and imagine a heavyweight wrestler lying face down and spread-eagle on the mat. Every time the opponent grabs hold of an arm or leg to turn him over, the wrestler immediately shifts his body so that the attacker never gets into a position that gives him good leverage (Chaabene et al. 2017). This situation could be compared with trying to use a crowbar to raise a rock and finding that the rock keeps moving, never letting you get into a position where you can use the crowbar to your advantage.

Compare the heavyweight wrestler's situation with that of a gymnast balancing on one foot on the beam (see figure 8.5). The wrestler does everything possible to maximize his rotary stability. He is massive, his center of gravity cannot be positioned any lower to the ground, and he constantly shifts his mass and spreads his arms and legs to counteract the torque applied by his attacker. The gymnast's center of gravity is high above the beam, and to make matters worse, she's balancing on one foot! Both the gymnast and the wrestler are successfully

maintaining their stability. But the gymnast has minimal rotary stability because little torque is needed to destabilize her. The wrestler, on the other hand, constantly repositions himself to maintain a tremendous amount of rotary stability.

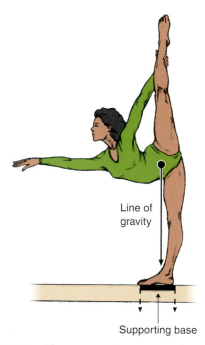

FIGURE 8.5   The gymnast's supporting base is the area covered by her foot on the beam. The line of gravity falls within the area of the base.

AT A GLANCE

Stability and Instability

- Stability is related to mass and inertia, in particular, how the preservation of an athlete's stability is always a battle of torques.
- The more stable a human is, the more resistance she can generate against any disruptive forces.
- Linear stability is concerned with an athlete who is moving in a particular direction; rotary stability is concerned with an athlete who is rotating.

Factors That Increase Stability

The conditions that give the gymnast minimal stability and the wrestler a considerable amount of stability are clues to the mechanical principles that determine different levels of stability. These

APPLICATION TO SPORT

Stability and Throwing

When objects or athletes rotate, swing, or turn, their rotary stability depends on their rotary resistance (i.e., rotary inertia). The greater the rotary resistance is, the greater the stability is. From chapter 6 you'll remember that rotary resistance is made greater by increasing the mass of the object and by moving the mass as far away as possible from the axis of rotation. How does this type of rotary stability show up in sport? Here is an example: The men's discus at 2 kg (4 lb 6 oz) is twice as massive as the women's discus, and it's over 5 cm (2 in.) larger in diameter. If all other factors are equal, the men's discus (by virtue of its greater mass and mass distribution) will battle the destabilizing effect of air currents better than the women's discus will.

Similar to a discus, a football will remain more stable in flight when it is given spin. The more spin it has, the greater its angular momentum and stability are. In this situation the angular momentum is increased simply by increasing the ball's angular velocity, or spin. Its mass or the distribution of its mass has not changed. When a quarterback rifles a pass to a receiver, the ball simulates the flight characteristics of a bullet. The spin around the ball's long axis gives the ball gyroscopic stability (i.e., stability resulting from spin). This stability helps the ball resist destabilizing forces produced by air currents and air resistance. Without spin, the ball will tumble and flutter in the air. Its flight will not be true, and it will not travel as far. The same thing happens when a discus is thrown without spin.

principles occur in every sport skill. Let's look at them one at a time to see their effect on how a human being will move or function.

Increase the Size of Base of Support

The bigger an athlete's base of support is, the greater the athlete's stability is. **Base of support** commonly refers to the area on the ground enclosed by the points of contact with the athlete's body. But the base of support is not always below the athlete. Anything that provides resistance against forces exerted by the athlete can become a base.

- If a student in an aerobics class leans against the wall in a calf-stretching exercise, her base includes the wall and the ground.
- A gymnast hanging from one of the uneven bars has a bar for a base, and it happens to be overhead.

What do we mean by *area* when we talk about the base of support? Imagine a gymnast who is facing along the length of the beam and is on one foot. In this situation, she uses the area of her foot as a base of support. If she places her other foot on the beam, her base of support now stretches from one foot to the other. The gymnast is now more stable than when she is on one foot alone. As you can guess, with two feet on the beam, her stability is better forward and backward than it is from the side. Shoving the gymnast off the beam is more difficult if you push her from the front or the back than if you push her from the side. In figure 8.4, the defending wrestler's base of support stretches from the single hand on the mat to where he contacts his opponent. Because his opponent is trying to turn him over onto his shoulders for a pin, the opponent can hardly be considered part of a stable base of support!

You can see from these examples that stability is directly related to the size of the supporting base. An athlete who makes the base as big as possible is more stable. A gymnast in a one-handed handstand has a base that is solely the area of one hand. If the gymnast moved to a headstand, the base would become the triangular shape that runs from the head to the hands and back again (see figure 8.6).

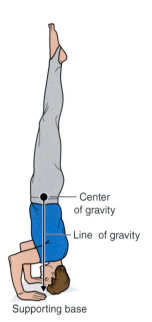

FIGURE 8.6 In a headstand, the supporting base is the triangular area from the head to the hands.

When a wrestler lies facedown and flat on the mat and spreads out his legs and arms as wide as possible, he covers a huge area that runs from his fingertips to his feet. As far as the size of his base is concerned, the wrestler has made himself maximally stable. Wrestlers make use of this position when both defending and attacking. If a wrestler is attacking and wants to hold the opponent in a particular position and stop him from moving, he maximizes his stability just as he would if he were defending. Maximizing stability makes it as difficult as possible for the opponent to mount a counterattack.

Centralize Line of Gravity Within Base of Support

An athlete's balance is maintained as long as a vertical line passing through the athlete's center of gravity falls inside the perimeter of the base of support, as shown in figure 8.5. The closer to the center of the base this line of gravity (i.e., a vertical line from the athlete's center of gravity) falls, the

more stable the athlete becomes. Conversely, the closer to the edge of the base the line of gravity falls, the more unstable the athlete becomes. The larger the base is, the easier it is for an athlete to make sure that this vertical line falls well within the base.

Unlike inanimate objects, living beings can maneuver and shift position and in this way keep their line of gravity within the perimeter of their supporting base. If you are balancing above a tiny base and allow your center of gravity to shift the slightest distance, then you've given gravity or an opponent a chance to apply a torque that you may not be able to counteract. For that reason, a gymnast's one-handed handstand is a phenomenal feat that requires tremendous strength and control. The base of support is solely the area covered by the gymnast's supporting hand. The center of gravity cannot be allowed to shift from directly above the supporting hand. So the muscles are put to work to maintain this position.

In total contrast to the one-handed handstand is the wrestler's spread-eagle defensive position. Assuming that the wrestler's center of gravity is close to his navel, the line of his center of gravity would have to be shifted a meter (several feet) in any direction before it got anywhere near the perimeter of his base. Not surprisingly, wrestlers often use this position.

You must not think that athletes always want to have their line of gravity within their base. When you walk or run, you take a series of steps forward. With each leg movement your center of gravity shifts outside your base and you shift from a stable position to an unstable position. When you put your foot down at the end of the step, your center of gravity moves back between your feet, and you are in a stable position again (see figure 8.7).

- When athletes sprint, they go from one unstable position to another as they drive themselves along the track.

- One moment their line of gravity is outside their supporting base, and the next instant it is back inside again. This sequence happens all the way down the track.

- The same situation occurs when hockey players skate along the ice.

When runners, speed skaters, cyclists, and slalom skiers go around a curve, they all lean into the turn. This position is an example of dynamic stability (i.e., stability while on the move). If they came to a sudden stop in the middle of the curve, they would fall. As they lean into the curve, their

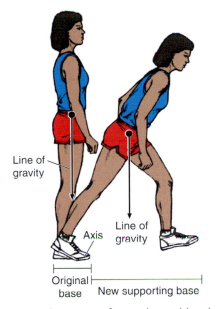

Line of gravity

Axis   Line of gravity

Original base   New supporting base

FIGURE 8.7 Taking a step forward, an athlete becomes unstable, restabilizing when both feet again contact the ground.

line of gravity no longer falls within their supporting base. The faster they move and the tighter the curve is, the more they lean inward and the farther

their center of gravity (and their line of gravity) shifts into the curve. The correct amount of lean balances the forces acting into the curve with those acting in the opposing direction (see figure 8.8).

We've been talking at length about maximizing stability. Do athletes perform some skills in which they purposely minimize their stability? Yes. Examples occur during the explosive acceleration of basketball or hockey players during a fast break and in swimmers and sprinters at the start of their races. In a sprint start, sprinters aim to get out of the blocks as fast as possible. On the "set" command, they shift their line of gravity forward so that it is close to their hand positions on the track (see figure 8.9). This highly unstable position satisfies two requirements:

- First, it extends the sprinters' legs into a powerful thrusting stance.
- Second, before the start, it moves the athletes as far as possible in the desired direction toward the finish line.

In sport skills that involve sudden direction changes, athletes want to be able to shift quickly in any direction. The set position in a sprint start is excellent for sudden, fast movement toward the

FIGURE 8.8 An athlete rounding a curve is in a state of dynamic balance. Forces acting into the curve counteract those acting outward.

Julian Finney/Getty Images

Line of gravity

FIGURE 8.9 In a sprint start, the athlete's line of gravity is shifted close to the forward edge of the supporting base.

finish in a 100 m race. But you'd be highly amused if you saw tennis players using a sprinter's set position when they're waiting to receive a serve. The set position might be good for a sudden move in one direction, but it is useless for moving quickly in other directions.

Volleyball players receiving a serve and soccer goalkeepers defending the goal all need to be quick off the mark in any direction. They cannot foresee the direction, velocity, and spin of the ball, and they certainly don't want to commit themselves too early. Consequently, these athletes use a small base with their line of gravity centralized. In this way they need no more than a split second to shift in the direction they want to move. They can react quickly and move fast in any direction.

Lower Center of Gravity

Athletes with a center of gravity that is elevated high off their base of support tend to be less stable than athletes whose center of gravity is lower. For this reason, great running backs who can twist and turn and suddenly cut one way and then the other tend to be shorter than other football players. They run low to the ground, so they are more stable and maintain their balance better than taller athletes can.

To understand how the degree of stability relates to the height of an athlete's center of gravity, let's have an athlete using the same-size base first stand erect and then crouch down. In both positions we'll tip the athlete sideways the same angle. Notice in figure 8.10 that for the same angle of tilt, the line of the athlete's center of gravity shifts beyond the edge of the supporting base when the athlete is standing erect. When the athlete is crouching down, as shown in figure 8.11, the line of gravity is still within the base. If other stability variables remain the same, the lower the athlete's center of gravity is, the more stable the athlete becomes.

The principle of lowering the center of gravity to increase stability is one of the reasons athletes crouch or lie flat on the mat when they are defending in combat sports. Athletes in wrestling and judo not only lower their center of gravity but also

---

APPLICATION TO SPORT

A Tightrope Walker's Pole Increases Stability

Circus tightrope walkers like the Great Wallendas and Philippe Petit of France have walked across cables strung between tall buildings. Why are these daredevil performers more stable when they use a long pole, and why are they even more stable when the pole is really long, curved downward, and weighted at either end? A pole that curves downward lowers the center of gravity of performer and pole combined. If weights are added at the ends of the pole, the pole lowers the center of gravity even farther. In addition, the longer the weighted pole is, the greater its resistance is against rotation. A long,

Daily Mirror/Mirrorpix/Mirrorpix via Getty Images

curved, weighted pole helps stop the performer from tipping sideways and falling off the wire. Without such a pole, tightrope walkers would have a tough time maintaining balance!

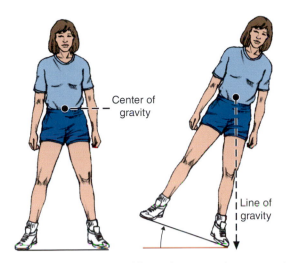

FIGURE 8.10 For an athlete who is standing erect, the line of the athlete's center of gravity shifts beyond the edge of the supporting base when the athlete is tilted.

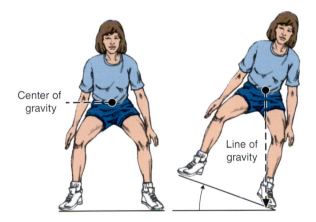

FIGURE 8.11 For an athlete who is crouching, the line of the athlete's center of gravity remains within the base when the athlete is tilted.

widen their base of support. This practice increases their stability twofold, making it considerably more difficult for an opponent to destabilize them.

Ski jumpers know that they lose style points if they stagger on landing. Lowering their center of gravity by using a telemark landing in which the legs are flexed helps them maintain their balance (Chardonnens et al. 2014). Kayakers who are tall and who have most of their weight in their upper bodies have a high center of gravity. They are less stable than shorter athletes or athletes who have most of their weight in the area of their hips and seats (taller kayakers can counteract this problem by using kayaks that have a wider beam, which widens their base of support).

Weightlifting gives us many great examples of reduction in stability when the center of gravity is elevated. The more weight a weightlifter hoists and the longer the athlete's arms are, the higher is the combined center of gravity of athlete and barbell. The line of this common center of gravity must stay centralized above a relatively long but narrow base formed by the lifter's feet. In addition, the barbell and the athlete's extended arms tend to rotate at the axes of the shoulder joints.

If an athlete allows the barbell to shift the slightest distance out of line, he must immediately reposition his base and fight with his shoulder muscles to bring the barbell back in line so that it is again centralized above his base. Figure 8.12 shows the final position in the jerk phase of a clean and jerk.

FIGURE 8.12 In a clean and jerk, the common center of gravity of the weightlifter and bar must be centralized above the supporting base. The barbell also must be positioned directly above the shoulder joints.

Stability in this position is better forward and backward because the base is large in that direction. But the base is narrow from side to side. Even small movements by the athlete in a sideways direction can shift the common line of gravity of athlete and barbell outside the periphery of the base. From the final leg split position of the jerk, the weightlifter must bring his feet together, stand erect, and control the barbell for 3 s. Obviously, controlling a heavy barbell in this position is difficult. The slightest shift of the barbell out of line can cause loss of control and failure of the lift.

Increase Body Mass

This principle simply says that if all other factors relating to stability remain unchanged, a heavier and more massive athlete is more stable than one with less body mass. For that reason, combat sports establish weight divisions. What hope would a featherweight wrestler have trying to lift and rotate an athlete in the super heavyweight division?

The value of body mass in combat sports is duplicated in American football, in which huge linemen are given the duty of maintaining their position in the face of whatever is thrown against them. The heavier they are, the more force (and torque) it takes to throw them off balance and knock them out of position (Brock et al. 2014). Therefore, they weigh close to or more than 136 kg (300 lb). But football has no weight divisions as there are in Olympic wrestling!

In all sport skills, heavier athletes who get out of control and lose their balance must exert more muscular force to regain their balance than lighter athletes do. If heavier athletes don't have the muscular strength to control their actions, their extra body mass becomes a great disadvantage and a severe liability.

In judo, athletes always try to make use of their opponent's body weight, and if their opponents are on the move, they try to make full use of their opponents' momentum. Trying to halt an opponent's push or thrust and then drive the athlete in the opposing direction is inefficient and exhausting. Better to use the opponent's movement and have him rotate around an axis (such as your hip or leg) and then add your force to that of gravity as the opponent rotates toward the mat.

Smaller sumo wrestlers who try to overcome heavier opponents also attempt to carry out these maneuvers. Obviously, shifting out of the way when an athlete weighing 181 kg (400 lb) is charging at you is a good idea. If you're a lighter opponent, you could try to destabilize him (by quickly stepping aside and simultaneously tripping him) as his huge mass rushes by. You then add your own force to that of gravity to drive him to the floor or, more satisfyingly, to heave him among the spectators!

Extend Base in the Direction of an Oncoming Force

Irrespective of the type of base, stability increases if an athlete's supporting base is enlarged in the direction in which force is being received or applied. For example, a force can come from an opponent who is trying to block or tackle you, or you can apply a force in a particular direction when you are throwing, hitting (e.g., in baseball), or lunging (as in fencing). When a running back wants to keep his stability and continue running when hit by an opponent, he must consider not only the force of the hit but also the direction it's coming from.

- To maintain stability and stay upright, the running back must widen his base in the direction of the applied force.
- If the hit is coming from the front, he widens his base from front to back.
- If the hit comes from the side, he widens his base in that direction.
- Naturally, he is going to lean into the hit as well. The mechanical principles of leaning into a hit are explained later in this chapter.

The second application of this principle involves situations in which athletes apply force in a particular direction. If you watch baseball players, you'll notice that they widen their base in the direction in which they are applying force, as shown in figure 8.13. They do this so that they have a good stable base that allows them to apply force over a considerable distance without losing balance. If they didn't do this, they'd be thrown in the opposite direction. Whether they are blasting a home run, hurling a fastball, or just having fun in a pickup softball game, the same principles apply.

The actual size of the base an athlete should use depends on how much force is applied. Imagine tossing a superlight table tennis ball a few meters. You could do this without any trouble while balancing on one foot. The table tennis ball has little mass and in this situation has hardly any velocity. Now try throwing a huge medicine ball as far as possible while balancing on one foot. You'll find that as the medicine ball goes in one direction, you get pushed in the opposite direction.

Let's now compare catching a table tennis ball that is lobbed gently toward you with catching a heavy object (like the medicine ball) traveling at a much higher velocity. You could easily catch the table tennis ball while balancing on one foot. Its momentum would be minimal, and you'd also be able to maintain your balance on one foot without any trouble. The medicine ball presents a different situation. If you want to maintain your stability and not have the medicine ball knock you over, you must assume a wide stance with both feet well planted on the ground. Reach toward the ball and extend your stance in the direction from which the

FIGURE 8.13   A baseball player lengthens his supporting base in the direction in which force is applied (i.e., in the direction of the pitch).

Harry How/Getty Images

medicine ball is approaching. The more massive the ball is and the greater its velocity is, the more stable your stance must be.

Shift Line of Gravity Toward an Oncoming Force

The principle of shifting your line of gravity (and your center of gravity) toward an oncoming force is directly related to the principle stating that you should widen your base toward an oncoming force.

- Running backs purposely lean into tackles.
- Hockey players lean into opposing players who are trying to body-check them.
- Wrestlers lean into their opponents when they grapple.
- Baseball batters lean into the oncoming pitch.

An interesting aspect of this principle is that athletes are temporarily destabilizing themselves by moving their line of gravity closer to the perimeter of their supporting base. This position requires the attacking athletes to apply considerable force (and momentum) to drive the defenders back beyond the rear perimeter of their supporting base. If the attackers are successful, then the defenders are destabilized. But the time taken by the attackers in thrusting the defenders from the forward perimeter of their base to beyond the rear perimeter is frequently sufficient for the defenders to spin away from the attack and to turn defense into attack.

The principle of shifting your line of gravity toward an oncoming force applies not only in tackles and checks, but also in catching a heavy medicine ball. You widen your base and shift your center of gravity toward the oncoming medicine ball. In this way the momentum of the medicine ball pushes you back into a stable position.

Equally important, you give yourself plenty of time to apply force to slow down the medicine ball. Do you recognize that "plenty of time to apply force" is an expression of impulse used for stopping (i.e., a small force applied over a long period)? A small force applied over a long period rather than a huge force applied over a short time allows you to bring the heavy medicine ball comfortably and painlessly to a stop.

An important difference exists when you are an athlete applying force. In this case you lengthen your base in the direction in which you are applying force. But you position your center of gravity to the rear of your base, in many cases temporarily outside the rear of your supporting base. All good javelin throwers start their throws from a position where they are leaning backward with their center of gravity outside the rear of their base of support; then in the follow-through, they finish with their center of gravity beyond the front edge of their base (see figure 8.14). In this way they apply great force over the longest possible distance and the longest possible time.

In combat sports like wrestling and judo, athletes must be aware of the danger of shifting their center of gravity too close to the perimeter of their base. Imagine that you're in a judo competition. You're pushing and leaning into your opponent, and the opponent is doing the same to you. Your center of gravity is close to the front edge of your base. Suddenly your opponent stops pushing and instead pulls hard. Now the force of your own push has suddenly been increased by the addition of your opponent's pull. You find that your center of gravity is suddenly outside your base and that you're unstable. You're about to be thrown!

If you sense that a push is about to change into a pull, immediately shift your center of gravity in the opposing direction and widen your supporting base in the same direction. Attack–counterattack sequences involving the positioning of the athletes' center of gravity are the essence of judo. One instant you can be stable, but if you are not careful, in the next instant you can be rotating toward the floor around an axis set up by a leg sweep from your opponent.

Remember that all the factors that control rotary stability are closely interrelated. Making one maneuver to increase stability is ineffective if all the other factors controlling stability aren't satisfied as well. For example, if an athlete widens her base of support, she must also move it in the direction of an applied force. A soccer player can widen her base toward a tackle coming from the right, but if her line of gravity stays close to the left edge of her base, she will easily be knocked over and beaten by the tackle.

AT A GLANCE

How to Increase Stability

An athlete can increase stability in several ways:

- Increase the size of base of support
- Centralize the line of gravity within base of support
- Lower the center of gravity
- Increase body mass
- Extend the base in the direction of an oncoming force
- Shift the line of gravity toward an oncoming force

FIGURE 8.14   In the javelin throw, the athlete's center of gravity begins to the rear of the supporting base and ends in front of the base.

A wrestler in the super heavyweight division is more stable than one in a lower weight division, but extra body weight is of little use if this super heavyweight uses a narrow base and stands with his center of gravity high off the ground and close to the edge of his base. Likewise, an athlete who lowers his center of gravity improves his stability, but this action is a valuable maneuver only if he keeps his line of gravity well within his base. All the principles of stability are related, and each depends on the others. Obeying only one principle of stability isn't good enough.

How to Measure Center of Gravity and Line of Gravity for an Athlete

The center of gravity is sometimes described as the point about which a body would balance without a tendency to rotate. Measuring the center of gravity and line of gravity requires only simple technology. To determine this balance point for a sporting implement, we can simply use an angle, about 0.3 m (1 ft) long, with the width of the angle in each direction about 5 cm (2 in.), just like a seesaw in a children's playground.

- If you place the angle on a flat surface with the apex pointing up, like a triangle, this edge acts as a pivot.

- To determine the center of gravity for an implement, you simply place the object on the apex. By slowly moving the implement around on this apex, you can determine its balance point, or center of gravity.

- Placing a mark on the sport implement allows you to see where the balance point is located.

- An example of the center of gravity for several sport implements is shown in figure 8.15.

- If you want to measure the line of gravity, you can use a simple string line and plumb bob, such as those used in building construction. The mass hanging on the bottom of the plumb bob will always hang vertically because of gravity, so this device allows you to identify the line of gravity.

For example, imagine a football player who moves his feet in the direction of the opposition to lengthen his base of support and improve his stability before a tackle. To show where the football player's line of gravity is located, and thus where he would position his body and feet to be most stable, you can hold a plumb bob from his belly button to show where the line of gravity will fall. He can then alter the position of his feet, and the difference in the line of gravity can be measured to determine what positioning would provide the most stability.

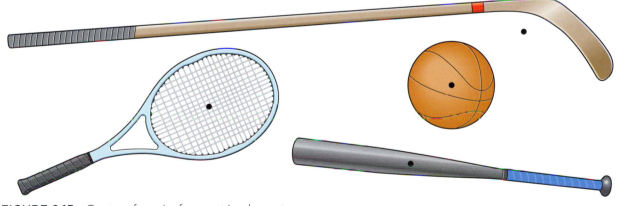

FIGURE 8.15   Center of gravity for sport implements.
Adapted from S.J. Hall, *Basic Biomechanics*, 4th ed. (Boston: McGraw-Hill, 2003).

SUMMARY

- To compete in sport an athlete must be stable. Stability refers to an object's (or athlete's) resistance to having its balance disturbed. Stability differs by degree. An athlete can be highly stable or minimally stable.

- From mechanical principles, there are two types of stability: linear and rotary.

- Linear stability can exist when an athlete or an object is at rest or moving in a straight line. At rest, linear stability is proportional to an athlete's mass and the frictional forces that exist between the athlete and any supporting surfaces. When an athlete is moving, linear stability is directly related to mass and inertia. The more massive the athlete is, the greater the inertia is and the greater the linear stability is.

- The rotary stability of athletes or objects is their resistance against being tipped over, or upended, and their resistance against being stopped or slowed down when they are rotating.

- Rotary stability is a battle of torques. The turning effect of a torque that disturbs an athlete's balance must be countered by the turning effect of a torque that the athlete applies to regain balance.

- Several principles relate to an athlete's stability. The athlete should satisfy as many of these principles as possible to ensure that stability is maximized: (a) Stability is increased when an athlete's area of supporting base is made larger; (b) stability is increased when the athlete's line of gravity falls within the boundaries of the athlete's supporting base, especially when the athlete's line of gravity is centralized within the supporting base; (c) stability is increased when an athlete lowers her center of gravity; (d) stability is proportional to an athlete's mass—the greater the athlete's mass is, the greater the stability is; (e) to maintain stability after impact, an athlete's supporting base should be extended toward an oncoming force; and (f) to maintain stability after impact, the athlete's center of gravity and line of gravity should be shifted toward the oncoming force.

- Some sport skills require minimal stability. Sprinters and swimmers in a set position shift their line of gravity toward the edge of their base in the direction in which they will race. Athletes who need to move quickly in any direction keep their base small and centralize their line of gravity.

- In a situation in which objects or athletes are rotating, rotary stability is proportional to angular momentum. Rotary stability is increased by increasing mass, increasing angular velocity, and increasing the distribution of mass away from the axis of rotation.

KEY TERMS

balance	linear stability
base of support	rotary stability
line of gravity	stability

REFERENCES

Barbado, D., A. Lopez-Valenciano, C. Juan-Recio, C. Montero-Carretero, J.H. van Dieën, and F.J. Vera-Garcia (2016). "Trunk Stability, Trunk Strength and Sport Performance Level in Judo." *PLoS One* 11 (5): e0156267.

Brock, E., S. Zhang, C. Milner, X. Liu, J.T. Brosnan, and J.C. Sorochan. 2014. "Effects of Two Football Stud Configurations on Biomechanical Characteristics of Single-Leg Landing and Cutting Movements on Infilled Synthetic Turf." *Sports Biomechanics* 13 (4): 362-79.

Chaabene, H., Y. Negra, R. Bouguezzi, B. Mkaouer, E. Franchini, U. Julio, and Y. Hachana. 2017. "Physical and Physiological Attributes of Wrestlers: An Update." *Journal of Strength and Conditioning Research* 31 (5): 1411-42.

Chardonnens, J., J. Favre, F. Cuendet, G. Gremion, and K. Aminian. 2014. "Measurement of the Dynamics in Ski Jumping Using a Wearable Inertial Sensor-Based System." *Journal of Sports Sciences* 32 (6): 591-600.

Falda-Buscaiot, T., F. Hintzy, P. Rougier, Patrick Lacouture, and N. Coulmy. 2017. "Influence of Slope Steepness, Foot Position and Turn Phase on Plantar Pressure Distribution During Giant Slalom Alpine Ski Racing." *PLoS One* 12 (5): e0176975.

Punt, M., S.M. Bruijn, H. Wittink, and J.H. van Dieën. 2015. "Effect of Arm Swing Strategy on Local Dynamic Stability of Human Gait." *Gait and Posture* 41 (2): 504-509.

 Visit the web resource for review questions and practical activities for the chapter.

9

Sport Kinetics

Vaughn Ridley/Getty Images

When you finish reading this chapter, you should be able to explain

- what is meant by work, power, and energy in sport;
- the relationship and differences between kinetic energy, gravitational potential energy, and strain energy;
- how momentum is conserved and how kinetic energy is dissipated; and
- how the law of conservation of energy can be applied in sport.

So far in this text on applied sport mechanics we have applied Newton's three laws of motion to describe and measure motion of an athlete or a sport implement. The analysis of human motion can also be conducted from the perspective of energy and work, that is, the influence work has on the energy of an athlete or object. This chapter expands on Newton's laws and introduces the standard physics measures of work, power, and energy. As in previous chapters, the application of these components to sport is paramount. We begin by exploring the fundamental principles of sport kinetics and then apply this knowledge in a sport perspective. By using this objective process we can explain what is happening in sport scenarios.

The mechanical terms of *work, power, energy, rebound,* and *friction* are often used in everyday situations, such as, "I've been working hard training in the gym." This section of the text quantifies and clarifies what we mean when using these descriptors. When we move, our movement can involve the following fundamental mechanical concepts, each of which is discussed in the following sections:

- Work (a measure of how much force has been applied over a certain distance)
- Power (a measure of how much work has been done over a certain period)
- Energy (a measure of how much work we can do)
- Rebound (a measure of the spring back after hitting or colliding with something)
- Friction (a measure of the resistance to movement)

Work

In everyday use, **work** usually describes some kind of activity not as pleasurable as play. Working out in the weight room, although it can be fun, suggests labor and hard work. In applied sport mechanics, when work is done, it specifically means that an object or athlete has applied force over a particular distance. Because distance is involved, the object or athlete moves from one position to another. From the earlier chapters we know that force applied over a particular distance must be related to **impulse**, whereby force is applied over a certain length of time. In a mechanical sense, work and impulse are related when a resistance is moved over a certain distance.

An athlete applies force to the javelin for a certain amount of time and over a particular distance. So in applying an impulse to the javelin, the athlete is also performing work on the javelin (Stefani 2014). This impulse can also resist movement. For example, after the player hits the ball in field hockey, the resistance from the grass will slowly bring the field hockey ball to a stop; the grass progressively applies force to the ball. This example is an expression of the impulse of stopping, and it is also an example of mechanical work being done. In the two cases, force is applied to the javelin and the field hockey ball over a long period as well as over a large distance.

In some situations in sport, although the athlete is exerting herself, she technically is not producing any work. For example, during isometric exercises, muscular force is applied for a certain amount of time against a static, immovable object. A common exercise for an athlete is to push against a weightlifting bar positioned at head height in a rack for 10 s.

- If the bar doesn't flex or move, then no mechanical work is done.
- No matter how vigorously the athlete's muscles contract or how much physiological work is done, if a resistance is not moved over a certain distance, no mechanical work is done.

Power

Power refers to the amount of mechanical work done in a particular period. In everyday life we use **horsepower** as a measure of the power of machines and engines such as those involved in the Indy 500.

- One horsepower is the ability of a machine (or a human) to move 250 kg (550 lb) a distance of 0.3 m (1 ft) in 1 s.
- In the metric system, power is measured in watts (746 watts = 1 horsepower).

How is power used in athletic contests? Using a weightlifting example, imagine two athletes lifting barbells of the same weight. One takes 2 s to lift the barbell overhead, and the other takes 1 s. They lift the barbell the same distance. In this comparison, the latter athlete is more powerful. Why?

- The two athletes moved the same weight over the same distance and performed similar amounts of mechanical work.
- But the second athlete took less time and therefore is considered more powerful.
- In the majority of cases of training for sport, because power is the rate of force produced over time, power is a more important component than strength—which is simply how much mass can be moved, regardless of how fast.

Here's another example illustrating power. Let's imagine that two athletes (we'll call them Scott and Rick) race against each other over 100 m. They cross the line in a dead heat in 10.0 s. And let's say that on the day of the race, Scott is more massive than Rick. Scott is therefore more powerful because he moved more mass than Rick did over the same distance (100 m) in the same time (10.0 s). This scenario also indicates that power differs from strength (the ability of a muscle to exert force) because strength does not necessarily imply the application of force with speed. Perhaps on this basis, the sport of powerlifting, which tends to measure strength rather than power in squats, the bench press, and the deadlift, should be called strength lifting.

In many sports, power is tremendously important because a slow application of force will not get the job done, particularly in throwing and jumping events; in the snatch and the clean and jerk in Olympic weightlifting; and in sports like gymnastics, in which skills such as back and front somersaults cannot be performed slowly (MacKenzie et al. 2014). Successful performance in these events demands that great force be applied quickly over a particular distance.

Energy

In everyday life, an energetic person is one who has the capacity for action. In mechanics, **energy** specifically means the ability of an athlete or an object to do mechanical work (i.e., to apply force over a distance against a resistance). Mechanical energy comes in three forms:

- Kinetic energy
- Gravitational potential energy
- Strain energy

Kinetic Energy

The word *kinetic* means that motion is involved, and ***kinetic energy*** is the capacity of an object or an athlete to do mechanical work by virtue of being on the move. The more mass that objects or athletes have, and in particular the faster they move, the greater their capacity is to do mechanical work (Mytton et al. 2013).

How Momentum and Kinetic Energy Are Related

Any object or athlete who is moving will always have both momentum and kinetic energy. These two concepts are interrelated.

- All objects on the move have both momentum and kinetic energy.
- The more mass they have and the faster they move, the greater their capacity is to apply force over time (which we call impulse).

Isaac Newton's third law of action and reaction is always involved in these situations. It tells us that the impulse applied against another object or another athlete is reflected back against the person or thing applying the impulse. Let's look at momentum and kinetic energy in **impact** situations (i.e., collisions and tackles) to see how the concepts of momentum and kinetic energy are related and how they differ.

The easiest approach is to regard kinetic energy as the ability of a moving object to do work on whatever it collides with and in return to do work on itself. The moving object can be anything at all. It can be an athlete running into an opponent or an archer's arrow piercing a target. A moving object has the ability to do mechanical work, and it can apply force over a particular distance to whatever it hits. Kinetic energy is involved when an athlete drives into the opponent, a pole-vaulter compresses the landing pads, and an arrow buries itself in a target. The formula for kinetic energy is

$$\text{kinetic energy} = 1/2 \ m{\cdot}v^2$$

where m is the mass of the object and v is its velocity.

As you can see, kinetic energy is directly proportional to any increase in mass, but more important, it increases according to the square of the velocity. These characteristics indicate the following:

- If you leave the mass of a moving object unchanged but double its velocity, the kinetic energy of the object increases fourfold.

- If you leave the mass of a moving object unchanged but triple its velocity, its kinetic energy increases to nine times the original. Increasing the kinetic energy of an object ninefold gives it the ability to do nine times more mechanical work on whatever it hits.

Take the archer's arrow as an example: An archer fires an arrow at a target. The arrow pushes in a certain distance. The archer then picks an arrow that is twice as massive and hits the target at the same velocity as the original arrow. The result is that it pushes in approximately double the depth of the original arrow. When the archer goes back to the original arrow but shoots it so that it's traveling twice as fast when it hits the target, the arrow pushes into the target approximately four times deeper. An arrow traveling three times as fast would drive it in approximately nine times deeper. In each example, we need to use the word *approximately* because not all the arrow's kinetic energy is used in driving its head into the target. Some of the arrow's kinetic energy is dissipated as vibrations and noise, and some as heat, which slightly warms the arrowhead and the pierced target area.

An athlete on the move has both momentum and kinetic energy. Let's consider tackling in American football to see that momentum and kinetic energy are always related but at the same time quite different. Imagine a 136 kg (300 lb) lineman, as shown in figure 9.1*a*, entering a tackle at a velocity of 1.2 m/s (4 ft/s).

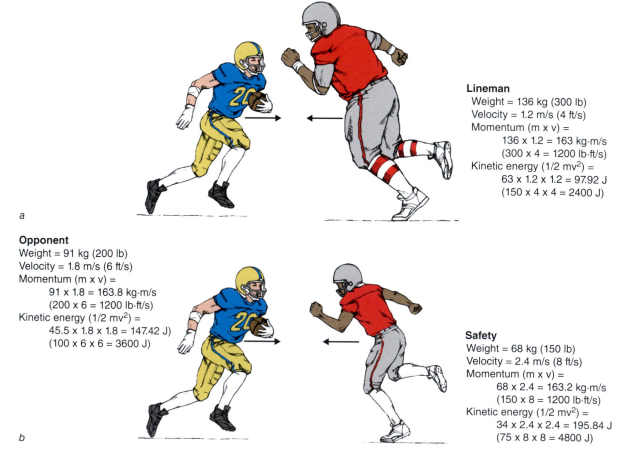

Lineman
Weight = 136 kg (300 lb)
Velocity = 1.2 m/s (4 ft/s)
Momentum (m x v) =
    136 x 1.2 = 163 kg·m/s
    (300 x 4 = 1200 lb·ft/s)
Kinetic energy (1/2 mv²) =
    63 x 1.2 x 1.2 = 97.92 J
    (150 x 4 x 4 = 2400 J)

Opponent
Weight = 91 kg (200 lb)
Velocity = 1.8 m/s (6 ft/s)
Momentum (m x v) =
    91 x 1.8 = 163.8 kg·m/s
    (200 x 6 = 1200 lb·ft/s)
Kinetic energy (1/2 mv²) =
    45.5 x 1.8 x 1.8 = 147.42 J)
    (100 x 6 x 6 = 3600 J)

Safety
Weight = 68 kg (150 lb)
Velocity = 2.4 m/s (8 ft/s)
Momentum (m x v) =
    68 x 2.4 = 163.2 kg·m/s
    (150 x 8 = 1200 lb·ft/s)
Kinetic energy (1/2 mv²) =
    34 x 2.4 x 2.4 = 195.84 J
    (75 x 8 x 8 = 4800 J)

FIGURE 9.1 Momentum and kinetic energy in a tackle. (*a*) The lineman and (*b*) the safety have the same momentum, but the safety has twice as much kinetic energy.

- At this velocity, the lineman has 136 × 1.2 = 163 units of momentum.

- A 68 kg (150 lb) safety, as shown in figure 9.1*b*, sprinting into a tackle at 2.4 m/s (8 ft/s) has the same amount of momentum (i.e., 68 × 2.4 = 163 kg·m/s, or 1,200 lb·ft/s).

- If each of their opponents weighs 91 kg (200 lb) and runs at 1.8 m/s (6 ft/s), then the opponents also have 163 kg·m/s (1,200 lb·ft/s) units of momentum.

- In the tackle, the 136 kg (300 lb) lineman and the 68 kg (150 lb) safety are equally effective in stopping their 91 kg (200 lb) opponents.

- In each case, 163 kg·m/s (1,200 lb·ft/s) units of momentum run into 163 kg·m/s units of momentum.

- If no rebound occurs and they stick together like clay, then in both tackles the athletes come to a stop on the spot where they collide.

Let's now look at the kinetic energy brought into the tackle by the lineman and the safety. Because the safety sprints at 2.4 m/s (8 ft/s), he goes into the tackle twice as fast as the lineman, who enters his tackle at 1.2 m/s (4 ft/s).

- If the lineman and the safety had the same mass, the safety would have four times the kinetic energy of the lineman because he runs twice as fast (remember, doubling the velocity increases kinetic energy fourfold [i.e., 2 × 2 = 4]).

- But at 68 kg (150 lb), he has half the mass of the 136 kg lineman. So his kinetic energy is 4 divided by 2, or twice as much as that of the lineman.

How is double the amount of kinetic energy expressed in the tackle? Think of the archer's arrow piercing the target or an athlete flexing the pole at the start of a vault. In his tackle, the safety is going to do twice as much work on his opponent as the lineman is. Think of him driving himself twice as far (just like the archer's arrow) into his opponent's body. The result of this extra ability to do work causes a lot of pain and maybe broken bones for his 136 kg opponent and for the 68 kg safety as well, because Newton says, "For every action, there is an equal and opposite reaction."

Kinetic energy figures in all situations in which something is on the move. Perhaps the most important lesson to be learned about kinetic energy is its effect when your car skids. A skid is an example of mechanical work being done by an

---

APPLICATION TO SPORT

How Technology Changes the Energy in the Tennis Racket

When a tennis racket connects with the tennis ball, the racket absorbs and then releases energy to the ball. The sweet spot is that part of the tennis racket that returns the ball with the greatest velocity and the least shock and vibration to the player. Because of advances in technology, tennis rackets have increased in size from 483 sq cm (75 sq in.) for the old wooden rackets to 613 to 645 sq cm (95 to 100 sq in.) for modern composites. On wooden rackets, the sweet spot was close to the base of the racket face. Modern composite rackets have a larger sweet spot, which is higher on the racket face. In addition, composite rackets are stiffer so that they transfer more energy to the ball. With a higher sweet spot, players hit the ball more reliably, and feel less shock, and the ball comes off the racket faster.

JACQUES DEMARTHON/AFP/Getty Images

object (you and your car). So imagine that you're driving your car at 16 km/h (10 mph) and suddenly have to slam on your brakes. And let's say that you skid 1.5 m (5 ft). What would be the approximate distance of your skid if you were traveling at 32 km/h (20 mph)? In this situation you've doubled your velocity, which squares its effect on the car's kinetic energy and its ability to do mechanical work (in the skid).

- Consequently, your car's stopping distance will be 6 m (20 ft), that is, four times longer than the original distance of 1.5 m (5 ft).

- An increase in the car's velocity to 48.2 km/h (30 mph) gives you $3 \times 3 \times 1.5 = 13.5$ m (44 ft).

- An increase to 64 km/h (40 mph) produces $4 \times 4 \times 1.5 = 24$ m (79 ft).

All these distances are approximate because kinetic energy is lost in generating heat and noise and because we estimated the length of the original skid and assumed that all other conditions stayed the same throughout. But an increase in velocity dramatically increases the kinetic energy of an object and its ability to do mechanical work. So the faster you drive your car, the more important it is to leave larger and larger distances between you and the car ahead of you and, if possible, to drive a car with antilock brakes, which minimize skidding.

How Kinetic Energy Is Dispersed

Now let's consider two football players, Jack and Pete. Imagine Pete running with the football and Jack approaching to tackle him. They are both moving, so they both possess momentum and kinetic energy. Is kinetic energy conserved by the two players in the tackle in the same way that momentum is?

No, because the kinetic energy that the two players bring into the tackle doesn't remain totally with the players but is spread around, or dispersed, in many ways. Some kinetic energy is certainly used by Jack in doing work on Pete and likewise by Pete in doing work on Jack. They squash into each other and maybe they rebound as well. Their kinetic energy does the squashing, and the elasticity of their bodies (strain energy, discussed later) can cause them to bounce apart.

But kinetic energy is also used in generating noise and heat during the tackle. Noise and heat occur in all sport collisions, such as when you hear

- the crack of bat on ball in baseball,

- the sound of ball against ball in the game of pool, and

- the slap of hand against ball in a volleyball serve.

The two colliding objects also get a bit warmer. Noise and heat always take some of the kinetic energy of two colliding objects. Consequently, in a collision, momentum is conserved by the colliding objects, which is often described as the **dissipation of kinetic energy**. In car accidents, metal is bent and crushed, and tremendous noise and heat are generated by the kinetic energy of the colliding vehicles. Most modern cars now have built-in "crushability." They are designed to dissipate the kinetic energy involved in the impact with minimal effect on the occupants. The same design is used for high-speed race cars.

Conservation of Linear Momentum

Momentum occurs any time an athlete or an object moves, and it plays a particularly important role in sport and in situations in which collisions occur. An easy way to think of momentum is to see it as a weapon that an athlete can use to cause an effect on another object or an opponent. A puck hit with immense velocity by a hockey player can have enough momentum to drive a goalie backward. Why?

Momentum is a function of mass and velocity, and although a puck is light and has little mass, National Hockey League players can hit it over 160 km/h (100 mph). When the puck hits the goalie, the puck and the goalie (plus pads, gloves, mask, skates, and stick) become a combined mass for an instant. The puck slows and loses some of its momentum. Because the goalie is driven backward, the goalie gains momentum in that direction. A similar example occurs when a tennis ball is served at such velocity that it knocks the opponent's racket backward. This process is due to what is called the **conservation of linear momentum**.

When two (or more) objects interact, such as when a baseball is struck by a bat or two hockey players slam into each other, the total amount of linear momentum of the two objects or players after the collision will be the same as the total amount that existed beforehand.

- If two football players bring a total of 100 units of linear momentum into a tackle when they collide with each other, then after the tackle, 100 units of linear momentum will still exist.

- Linear momentum is not gained or, surprisingly, lost; we say that it is conserved.

- The law of conservation of linear momentum is directly related to Newton's third law, which says that every action has an equal and opposite reaction.

Using two American football players from the earlier discussion, Jack and Pete, let's see how the law of conservation of linear momentum works:

1. If Jack tackles Pete in a game, Jack and Pete exert equal and opposite forces on each other during the tackle for the same period. If Jack applies a force against Pete for 1 s during the tackle, then Pete does exactly the same in return to Jack. It may not look this way when you watch a receiver being hit at the instant he receives the ball, but mechanically, this is what occurs.

2. From this explanation, we can see that Jack and Pete each experience equal and opposite impulses. Equal and opposite impulses means that the product of Jack's force multiplied by the time that he applies his force in the tackle is applied back to him in the opposing direction by Pete [F × t (Jack) = F × t (Pete)].

3. Because the impulses that Jack and Pete apply to each other are equal in amount and in opposing directions, the change in linear momentum of both Jack and Pete must also be equal and opposite. As an example, if Pete's linear momentum is increased by a certain amount during the tackle, Jack's linear momentum must be decreased by the same amount.

4. The combined linear momentum of Jack and Pete has not changed in any way as a result of the tackle. Linear momentum has been transferred from one athlete to the other, but no gain or loss has occurred in the total linear momentum of the two players. Consequently, we say that total linear momentum of the two players has been conserved.

Collisions occur throughout sport, and it makes no difference when one object is moving (e.g., a golf club) and the other (a golf ball sitting on a tee) is not moving. The ball gains linear momentum, and the club loses some of the linear momentum it had before impact. Collisions cannot create or dissipate linear momentum. Other examples include hockey players, who can skate at tremendous speed and therefore generate great momentum. They demonstrate their momentum in the bone-rattling body checks that characterize their sport. Likewise, offensive and defensive linemen in the National Football League, with their immense size and great acceleration over 40 yd (36.6 m),

generate tremendous momentum. Like the hockey player, the lineman who has the most momentum at impact is likely to dominate in a collision with an opponent. In any of these collisions in sport, all that happens is a transfer of linear momentum from one object to another.

Strain Energy

Strain energy, like gravitational potential energy, is a form of stored energy. Objects have the capacity to store strain energy if they have the ability to restore themselves back to their original shape after being squashed, pulled, twisted, or pushed out of their original (resting) shape—just like a spring.

Mechanical work must be done to put an object in this condition. After the object is distorted, its ability to regain its original shape quickly is a measure of its strain energy. An archery bow springing back to its original shape after being flexed is an example of strain energy. Using high-speed video, the deformation of a golf ball can be observed; the strain energy causes the ball to spring back to its original shape after being squashed by a driver (Tao 2013). Likewise the leaf and coil springs used on cars and motorcycles spring back to their original shape after being flexed. All are examples of strain energy.

Gravitational Potential Energy

As the name suggests, **gravitational potential energy** is a form of stored energy that results from the acceleration of gravity—an energy that is potentially available and ready to be put to use. Objects and athletes have gravitational potential energy when they are raised above the earth's surface and are primarily influenced by the gravity of the earth.

Because the sports we are considering take place on or close to the earth's surface, gravitational potential energy is the energy that objects and athletes have by being raised even a short distance above the earth's surface.

- The greater the height is and the greater their mass is, the more potential energy they have (height above the surface of the earth = distance; mass = resistance; and the force that accelerates them = gravity).

- Unlike objects that have kinetic energy, objects or athletes can have gravitational potential energy just by being positioned at a distance above the earth's surface.

- Objects or athletes don't necessarily need to be moving.

Mechanical work has to be performed against the pull of gravity to get athletes or objects up above the surface of the earth so that they have gravitational potential energy. Examples are athletes who climb the 10 m tower to somersault and twist on their way down to the pool. The athletes perform mechanical work climbing the ladder and raising their body mass upward. With every rung they climb, they increase their gravitational potential energy.

Roller-coasters are no different. Roller-coaster cars and their occupants are pulled by a machine to the top of the first hill on the track. At that point the car and its riders possess maximum potential gravitational energy for the height they've reached. When the roller-coaster car (or a diver) accelerates downward toward the surface of the earth, gravitational potential energy is converted primarily to the energy of motion (kinetic energy). We say "primarily" because not all the kinetic energy is expressed as motion. Some kinetic energy is converted into noise and heat.

FIGURE 9.2   In the pole vault, a flexed pole stores strain energy, which is later used by the vaulter.

> ## AT A GLANCE
> ### Work, Power, and Energy
> - The key measures of sport kinetics are work (a measure of how much force has been applied over a certain distance), power (a measure of how much work has been done over a certain period), and energy (a measure of how much work we can do).
> - Mechanical energy comes in three forms: kinetic energy, gravitational potential energy, and strain energy.
> - The word *kinetic* means that motion is involved. The more mass that objects or athletes have, and in particular the faster they move, the more kinetic energy is produced.

Kinetic Energy, Gravitational Potential Energy, and Strain Energy in Sport

Kinetic, gravitational potential, and strain energy are all well demonstrated in the pole vault. In this event, an athlete sprints flat out carrying the pole. The kinetic energy developed during the approach (because the athlete's mass and the mass of the pole are in motion) is used to flex the pole and load it with strain energy, as shown in figure 9.2.

When a vaulting pole flexes at takeoff, the athlete has performed mechanical work on the pole by applying force to it over a particular distance.

- If the athlete runs slowly during the approach, the pole is flexed less, so less strain energy is stored in it.
- Like an archer's bow that is pulled back only slightly, a vaulter's pole that is not flexed the optimal amount cannot do much work when it straightens out and drives the athlete skyward.
- Top-class pole-vaulters combine the skills of a gymnast and a sprinter.
- The faster they run and the taller and more powerful they are, the higher they can hold the pole.

A phenomenal athlete like Sergei Bubka of Ukraine, who has vaulted over 6 m (20 ft) both outdoors and indoors, is so powerful and such a good sprinter that he can hold extremely high on a long, stiff pole that is specially rated for his body weight of 80 kg (176 lb). If you can flex this kind of pole, you store a tremendous amount of strain energy in the pole, which is returned to you by rocketing you up toward the crossbar.

During the vault, the flexed pole straightens out and drives the athlete up toward the bar. The strain energy stored in the pole by the athlete now does mechanical work by propelling the athlete upward, and because he is rising in the air, he gains gravitational potential energy. At the high point of the vault the athlete rises no more. His kinetic energy is momentarily zero because, for an instant, he is not moving.

But the athlete is a long way above the surface of the earth and therefore has maximal gravitational potential energy for the height to which he has been raised. As the athlete accelerates toward the earth, gravitational potential energy is progressively replaced by kinetic energy. By the time the athlete hits the landing pad, he is moving at top speed for the distance fallen and therefore has plenty of kinetic energy.

The athlete depresses the landing pad (increasing the time and therefore reducing the impulse force), and as a consequence some of this energy will make the contacting surfaces slightly warmer. The depth at which he squashes into the landing pads, the noise of impact, and the heat that is developed are expressions of the work done by the athlete's kinetic energy. After the athlete is back on the surface of the earth and not moving, his gravitational potential energy is zero, and he has no kinetic energy either.

Many other examples of kinetic, gravitational potential, and strain energy occur in sport. An athlete bending a springboard in diving is an example of work being performed to the board to load it up with strain energy. In trampoline, which became an official Olympic sport in the 2000 Sydney Olympics, athletes use a pumping action to progressively increase the stretch in the springs of the trampoline. In this way, the stretched springs are given more and more strain energy. Strain energy then performs mechanical work by propelling these athletes high into the air. Trampolinists have zero kinetic energy when, for a brief instant, they are motionless at the top of their flight path.

As with pole-vaulters, their gravitational potential energy is maximal at the height to which the trampolinists have been raised. After they drop downward, their kinetic energy (which progressively increases during the fall) is spent primarily in stretching the springs of the trampoline bed. Some will be lost as noise and heat, but the trampolinists make up for this loss by repeating their muscular pumping action when they are in contact with the bed of the trampoline. Muscular thrust, coupled with the kinetic energy of the drop to the trampo-line bed, stretches the trampoline springs again for the next flight up in the air. When you compare trampolinists to divers, you can understand why trampolinists don't want to travel along the bed of the trampoline during their routines. They want to stay in the middle. Divers, on the other hand, must travel away from the board to avoid hitting it on the way down to the water.

AT A GLANCE

Energy in Sport

- Gravitational potential energy is a form of stored energy that results from the acceleration of gravity.
- Strain energy, like gravitational potential energy, is a form of stored energy.
- In a collision, momentum is conserved by the colliding objects, whereas their kinetic energy is not.

Law of Conservation of Energy

Although kinetic energy gets used up in the deformation of objects (and athletes) and into heat and sound, all that is really happening is that one form of energy is being changed into another, or the **Law of Conservation of Energy**. For example, imagine a trampolinist who makes a mistake in her routine, misses the trampoline, and falls hard onto a wooden floor. We know that as the trampolinist falls faster and faster, her gravitational potential energy is converted into kinetic energy and her kinetic energy is maximum just before impact. At impact, gravitational potential energy is zero. Because movement appears to have ended, it seems that kinetic energy has disappeared too. But has it really?

After impact, the kinetic energy of the trampolinist has been converted into the movement of millions of molecules, some expressed as noise, some as heat, some as the movement of the floor, and some as the deformation of the trampolinist's body.

- What has happened is that one form of energy—kinetic—has been exchanged for other forms of energy with no real loss.
- In reality, energy is conserved.

Rebound

When bats hit balls or balls bounce off floors, a collision, or impact, occurs. When the objects separate and one moves away from the other (or both move away from the other), we call it a **rebound**. What actually happens during and after the collision depends on many interacting factors. Let's look at what happens to balls when they are hit and as they bounce.

Often both the ball and the object it collides with are moving (e.g., a racket hitting a ball). Sometimes one object can be moving and the other can be momentarily stationary, as when a bowling ball hits a pin. After the collision the bowling ball slows down slightly and loses momentum. The pin accelerates and gains momentum.

In another common situation, a moving object (e.g., a squash ball or basketball) collides and rebounds from an immense stationary object such as a wall or the floor. In the previous examples, the angle at which the collision occurs can vary. The two objects can hit head on (for example, when a basketball is bounced straight down onto the floor), or they can glance off one another at an angle (as with a carom shot in pool or a bounce pass in basketball).

Rebound and Elastic Recoil

An important factor that determines what happens after a collision is the degree of elastic recoil that

objects have, meaning the force by which objects push back to their original shape. Some objects have little recoil, or **elasticity**, and they stick together like clay after colliding. A Hacky Sack footbag offers virtually no bounce at all. Likewise, an athlete who falls out of control onto the ground barely bounces back up into the air. The human body in this condition has little elasticity.

Golf balls, on the other hand, are well known for their elastic recoil. They become deformed (squashed out of shape) when hit by a club (Smith et al. 2015). Like a spring that is compressed, the ball stores strain energy, which is transformed into kinetic energy as it regains shape. Just like what happens with the pole-vaulter's pole, strain energy produces kinetic energy, which then makes a contribution toward the velocity of the golf ball as it comes off the club face.

Kinetic energy also contributes to the noise of impact and increases the warmth of the object (in this case a ball) slightly after it has been hit. The same thing happens to a baseball when it is hit. A baseball is flattened to almost half its size when it is hit for a home run. As the ball leaves the bat, it springs back to its original shape. The energy produced by this action adds velocity to the flight of the ball. This extra energy helps great baseball

---

APPLICATION TO SPORT

Coefficient of Restitution

The coefficient of restitution (COR) is a measure of the ability of an object like a ball to spring back to its original shape after being hit by a club, bat, or racket or after bouncing off a floor or a wall. It is essentially a measure of bounciness or resilience. All ball sports have specific rules controlling just how much bounce a ball is allowed to have. Rules also take into consideration the composition and manufacture of the club, bat, and racket (and strings). Too much or too little rebound can change the whole character of the sport. The highest COR rating is

MichaelMaggs Edit by Richard Bartz/Wikimedia

1, and the lowest is 0. Children's Super Balls and golf balls have tremendous bounciness and a COR between 0.8 and 0.9. Tennis balls have a COR of 0.72. Baseballs are required to have a COR between 0.51 and 0.57, and basketballs must have a COR between 0.76 and 0.80. Squash balls have a COR between 0.34 and 0.57 and are available in a wide range of rebound and hang-time ability. This continuum is intended to accommodate the various abilities of players. In addition, the COR of a ball can vary according to temperature; as the temperature rises, so does the amount of bounce.

hitters put the ball out of the park or into the upper row of the bleachers. The elastic recoil of any object also depends on the nature of the object with which it collides. A football can rebound with considerable velocity off artificial surfaces. This rebound differs dramatically from the rebound that occurs when the ball drops on wet, muddy turf.

Rebound and Temperature

Temperature affects the way that balls rebound after a collision. Heat causes the air inside a ball to expand, which increases the ball's ability to rebound. In squash, players spend time before a match rallying back and forth to heat up the ball so that it will bounce correctly. A cold squash ball is virtually a dead ball that has little or no bounce, making it much more difficult for players to return a shot. Elite squash players are required to play with a ball designed to bounce very little, even when it's warmed up and the air inside it is hot. This property challenges the players' ability. Novices, on the other hand, use a ball designed to bounce much higher after it's warmed up. The ball's additional time in flight gives novices more time to get into position and make their strokes.

Angle, Velocity, Spin, and Frictional Forces in a Rebound

Many factors determine what happens to a ball after it collides with another object. In addition to those items already mentioned, rebound depends on

- the ball's angle and velocity,
- whether the ball is spinning, and
- how much friction (discussed in more detail in the next section) occurs between the ball and whatever it is hitting.

These factors play a big part in table tennis. The rough, spongy surface of the paddle imparts tremendous spin on the ball. Even though the ball and the table have smooth surfaces, the effect of the spinning ball hitting the table at high velocity is dramatic. The direction of rotation of topspin makes the ball accelerate off the table at a low angle. Backspin causes the reverse to occur. Its spin is in opposition to the ball's movement, and the ball can slow down or even reverse direction.

The direction of spin plays a vital role in many sports. In billiards, the spin (or "English") given to the cue ball affects what happens to the ball that the cue ball strikes and what happens to the cue ball afterward. Similarly, backspin applied to a golf ball by the angle, or loft, of the club face helps give the ball lift and enables it to stay in the air longer (you'll read more about lift in the chapter 10). Backspin can also prevent a golf ball from rolling off the far side of the green.

- In basketball, good shooters give the ball backspin at the instant of release.
- They do this by making the ball roll off their fingertips as it leaves their hand.
- A basketball shot with backspin that hits the rim or backboard not only loses velocity when it hits but also is more likely to drop into the basket because of the backward rotation.

Rick Barry, one of the greatest foul shooters ever to play basketball (90% average), was well known for his underhand shooting technique. He could throw the ball with considerable backspin in this manner, and the backspin was instrumental in rotating the ball down into the basket.

Most modern basketball players shoot fouls the same way they shoot during the game. Even though they shoot overhand, the finger flick applied at the end of their release produces backspin. The backspin then helps sink the ball into the basket. Topspin, on the other hand, is more likely to cause the ball to rebound back out onto the court.

Friction

When we attempt to move or remain stationary, a contact force called **friction** can resist movement. The force of friction is present when a jogger's running shoes make contact with the road and when a bowling ball rolls down the surface of the lane (Morio et al. 2015). In both of these examples, friction occurs between two solid (i.e., nonliquid) surfaces. But friction also results when a ski jumper or a javelin flies through air or when swimmers move through water. In these situations both the air and the water act as fluids, and they produce fluid friction. We'll examine fluid friction in chapter 10 and concentrate in this chapter on friction that occurs between solid surfaces.

The demands for friction vary dramatically from one sport to the next and from one set of environmental conditions to another. An athlete may want plenty of friction at one instant and only minimal friction at another.

- In cross-country skiing the correct choice of wax, relative to the snow conditions, allows sufficient friction for traction but not so much friction that an athlete cannot glide on the skis when necessary.

- When football players make sudden changes of direction, they depend on friction for good traction.

- Unfortunately, on artificial turf, friction between an athlete's shoe and the turf can sometimes be too good. During tackles the athlete's foot can be trapped in one position. The result can be torsion (i.e., twisting) injuries to the knee and ankle.

Another example is the sport of tennis. The international Grand Slam circuit for professional players covers three different surfaces: hard courts, grass, and clay.

- Hard courts vary in texture; more grit used in the top dressing increases the friction between the ball and the court surface. Increased friction slows down the ball as it rebounds off the surface.

- Grass provides little friction. The short grass at Wimbledon is recognized as a fast surface. This surface benefits power players who serve the ball at over 160 km/h (100 mph). Clay is the opposite of grass.

- Rough clay courts, such as those at Roland Garros Stadium in Paris, are extremely slow surfaces. These surfaces reduce the effectiveness of the serve-and-volley game of power players who dominate on faster grass and hard courts. On clay, defenders have more time to get to the ball and to counterattack. The strategy of clay court specialists is to hit passing shots over and past power players.

Thomas Muster of Austria was a king of the endurance battles that occurred on clay court surfaces. He built up a 46-3 record on clay, bringing his 1995-1996 total to an amazing 111-5. Unfortunately, he was not able to replicate this level of success when he played on other surfaces. More recently, Rafael Nadal of Spain has consistently demonstrated his ability on the clay court and was named the King of Clay because of his 10th title in 2017 at the French Open. The Queen of Clay would have to be Chris Evert, who won seven French Open titles.

Two types of friction occur between solid surfaces:

- The first is **static friction**, which exists between the contacting surfaces of two rest-

ing objects and provides the resistive force opposing the initiation of motion.

- The second is dynamic friction, which can be subdivided into two types of friction: **sliding friction**, a resistive force that develops when two objects slide and rub against each other, and **rolling friction**, which produces a resistive force when objects—such as balls and wheels—roll over a supporting or contacting surface.

The relationship of friction and the force applied is shown in figure 9.3. In the calculation of static friction (that is, the amount of friction when an object is stationary), the friction force is the static friction coefficient (μ_s) multiplied by the normal or downward reaction force R.

To calculate the dynamic friction force, the same formula is used but the static friction coefficient is replaced with the dynamic friction coefficient (μ_k).

As shown in figure 9.3, the static friction force increases to a point just before movement. At that point the dynamic friction force is less than the maximum static friction force. In sport, all three of these frictional forces can oppose the motion of a single object.

For example, a field hockey ball at rest resists movement initially because of the static friction existing between the ball and the turf. After a player hits the ball, it may simultaneously slide and roll, in which case both sliding and rolling friction resist its motion. But only rolling friction remains after the ball is fully rolling and no longer sliding.

Static and Dynamic Friction

Let's examine static and dynamic friction by looking at an American football player shoving a blocking

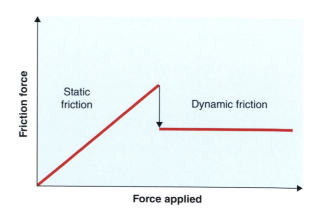

FIGURE 9.3   Relationship between friction and force.
Adapted from S.J. Hall, *Basic Biomechanics*, 4th ed. (Boston: McGraw-Hill, 2003), 390.

APPLICATION TO SPORT

Rebound in Tennis

The game of tennis is more complex than table tennis because of the number of factors that affect the rebound of the ball. For example, a tennis court can be clay, grass, or a hard surface. These surfaces have different effects on the rebound of the ball. Courts can be indoors or outdoors, and environmental conditions such as temperature, humidity, and wind velocity alter play. Tennis rackets vary in size, shape, weight, and flexibility. Strings differ in type and tension. Even the tennis ball itself can vary in the way that it reacts. Tennis balls bounce higher after they are warm and after some nap (i.e., fuzz) is worn off. When they have been out of their pressurized cans for a long time, balls age and lose their bounce irrespective of how much use they've had.

Allsport UK /Allsport/Getty Images   MANAN VATSYAYANA/AFP/Getty Images

PATRICK KOVARIK/AFP/Getty Images

sled, a common piece of equipment designed to slide on turf and provide both static and dynamic resistance against a player, as shown in figure 9.4. The frictional force generated by a blocking sled at rest comes from static friction. If an athlete pushes against the sled with minimal effort, the sled remains motionless.

In this situation the static friction is greater than the force produced by the athlete. If the athlete increases his thrust against the sled, the frictional force opposing the athlete reaches a critical level, called the sled's maximum static friction. Should the athlete further increase the force of his push,

he will overcome the sled's maximum static friction and the sled will start to slide and become dynamic.

Static friction is now replaced by a frictional force, called dynamic friction, between the base of the sled and the turf. Dynamic friction is always less than maximum static friction. In other words, keeping an object moving is easier than starting it moving. Now that we know the difference between static and dynamic friction, let's look at the factors that influence the amount of static and dynamic friction that can occur.

The relative motion between the base of the blocking sled and the ground refers to whether the

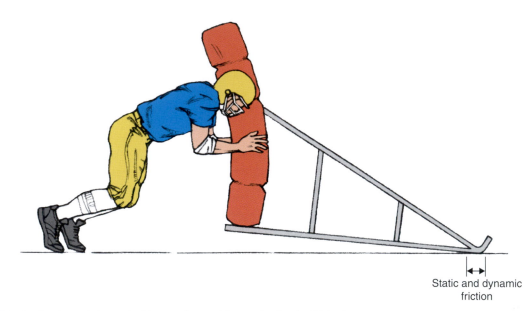

Static and dynamic friction

FIGURE 9.4   Static and dynamic friction during lifting of a football blocking sled.

athlete is applying force to initiate movement or to keep the sled moving. Remember that static friction produces more resistance than sliding friction. We all know from experience that keeping an object sliding is easier than starting it sliding.

Another example is a field hockey player who hits a ball, causing it to roll across the turf. Friction with the turf coupled with a small amount of air resistance slowly brings the ball to a halt. The small forces of friction and air resistance are applied to the ball over a long time (and distance), and the result is that they progressively reduce the velocity (and therefore the momentum) of the ball to zero. At any instant in time the force applied to the ball is small, but it is applied progressively over a long time.

Force Pressing Two Surfaces Together

Using the blocking sled as an example, the force pressing the base of the sled to the turf is equal to the weight of the sled pushing down and the reaction of the earth pushing up. These two forces press the contacting surfaces of sled and turf together. If a coach gets on the sled, both the weight of the sled pressing downward and the reaction force of the earth pressing upward increase. The frictional force opposing an athlete attempting to push the sled will be greater.

The importance of mass pressing two surfaces together cannot be overemphasized. More mass means that more pressure is thrusting the contacting surfaces of turf and sled together. For maximal friction, the mass of the sled plus whatever mass is

added must act perpendicularly to the supporting surface. If a coach hangs on the sled with his body angled (so that he is not standing upright), only part of his mass contributes to pushing the sled down onto the turf.

Actual Contact Area Between Two Surfaces

Actual is the important word here because friction can occur only where two surfaces are in contact. Where part of the base of the blocking sled does not contact the ground, friction cannot occur. If an athlete lifts the sled by pushing forward and upward, then no friction occurs where the sled loses contact with the ground. Likewise, if a soccer player's studs are the only part of the boot contacting the ground, then only the surface of the studs contributes to the contacting area; the rest of the sole makes no contact and generates no friction.

The following example illustrates the importance of union between the actual contact area (i.e., the parts of the surfaces in contact) and the forces pressing the two surfaces together. Imagine two sleds that weigh the same but have bases of different sizes.

- An athlete pushes against one sled and then in exactly the same manner (and with the same force) against the other.

- The sled with the larger base will be no more difficult to push than the sled with the smaller base, because the weight of the larger sled is spread over a larger area and presses onto the turf proportionally less for every square centimeter (inch) of its base.

- Consequently, the two sleds produce the same friction.

Nature and Type of Materials That Are in Contact

The nature and type of materials in contact have to do not only with the material on the base of the sled but also with the type of surface that the sled contacts. Imagine that the base of the sled is rough and pitted and that it is forced to slide over a muddy, sticky surface. The frictional force opposing the player would be greater with this sled than the force produced by a sled with a smooth base sliding over a hardened, level surface.

The many different materials used in sport (and everyday life) produce varying levels of friction as they contact one another. At one end of the scale are the long steel blades of a speed skater gliding on ice. A thin film of water created by the blade pressing on the ice produces a lubricant that reduces friction to an extremely low level. At the other end of the scale, court shoes with gum rubber soles pressing against a rubberized surface create an extremely high level of friction.

Rolling Friction

Rolling friction occurs when a round object, such as a ball or wheel, rolls across a contacting, or

AT A GLANCE

Conservation of Energy in Sport

- Although kinetic energy gets used up in the deformation of objects (and athletes) and into heat and sound, all that's really happening is that one form of energy is being changed into another.

- Temperature affects the conservation of energy, particularly in a rebound following a collision.

- When we attempt to move or remain stationary, a contact force called friction can resist movement.

supporting, surface. Rolling friction is common in bowling, billiards, golf, field hockey, cycling, and soccer. The resistive force generated by rolling friction is significantly less than that of sliding friction. Indeed, the use of the wheel is commonplace because of the minimal levels of friction produced as a wheel rolls over a supporting surface. This extremely low friction results from the ease with which a rounded surface detaches itself from the surface it is rolling over. Rolling friction varies according to the nature of the surfaces in contact, the pressure pushing the surfaces together, and the diameter of the rolling object.

APPLICATION TO SPORT

How a Rolling Wheel Makes Use of Static Friction

Many cars are equipped with antilock brakes that allow the wheels to continue to rotate while the brakes are progressively applied. This action is much better than what occurs when wheels suddenly lock and start to skid. A skidding wheel has less traction (grip on the road surface) than a wheel that is not skidding, and antilock brakes stop the car from skidding while slowing it down. Sensors in cars are designed to respond to a sudden reduction in rotation of the wheels. They continue to roll rather than lock up and cause a skid. With antilock brakes, the car stops faster and the driver can continue to steer. Steering is difficult when a car is skidding. A skidding tire is slowed by sliding friction between the contact points of each tire and the road surface. A rotating tire can be considered to have hundreds of contact points that are each stationary for an instant one after another. Therefore, a rotating tire is slowed down by static friction, which is more powerful than sliding friction.

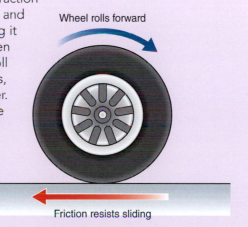

Wheel rolls forward

Friction resists sliding

APPLICATION TO SPORT

Work, Power, Energy, Rebound, and Friction in Sport

These terms are important in sport because they define and quantify the amount of movement that has occurred. For example, by knowing how to calculate work (force × distance), we can compare one athlete's performance to another's or, for one athlete, determine whether the training that the athlete has been doing is effective, as demonstrated by an increase in work. Similarly, the important variable of power (force × time) can be quantified, and the more powerful an athlete becomes, the more effective that athlete can be in a range of sports. Understanding

Roger T Schmidt/Photographer's Choice/Getty Images

the energy that is stored, generated, lost, and translated allows skillful mechanical intervention with the athlete's technique and, ultimately, enhancement of performance. Finally, knowledge about frictional forces can be used both to enhance a performance (e.g., increasing the friction between the ground and the shoes of a lineman will increase his ability to block an opposing player) and to make the sport safer (e.g., reducing the friction between the tennis court and the player will allow the player to slide rather than abruptly stop when running across the court). Most important, measuring these mechanical components will allow provision of objective feedback to the athlete and coach on what they should do in the current sporting activity.

Sprint cyclists use narrow tubular tires and inflate them to phenomenal pressures that average 827 kPa (kilopascals), or 120 psi (pounds per square inch). The result is that even with the weight of the cyclist, hardly any tire surface contacts the track, reducing rolling friction to a minimum. Even so, the rubber of the tire in contact with the smooth velodrome surface provides excellent traction. Rolling friction is the reason that fat-tire mountain bikes feel so sluggish on the road, particularly if their tires are underinflated. On the other hand, wide, knobby mountain bike tires are designed to produce great stability and traction on rough terrain. High-pressure, narrow tubular tires are virtually useless in those conditions.

SUMMARY

- Work is a measure of how much force has been applied over a certain distance.
- Power is a measure of how much work has been done over a certain period.
- Energy is a measure of how much work we can do.
- Rebound is a measure of the spring back after hitting or colliding with something.
- Friction is a measure of the resistance to movement.
- In most cases of training for sport, because power is the rate of force produced over time, power is a more important component than strength, which is simply how much mass can be moved, regardless of how fast.

- Mechanical energy comes in three forms: kinetic energy, gravitational potential energy, and strain energy.
- All objects on the move have both momentum and kinetic energy.
- The more mass that athletes or implements have and the faster they move, the greater their capacity is to apply force over time (which we call impulse).
- If the mass of a moving object is unchanged but the velocity doubles, the object's kinetic energy increases fourfold. The word *kinetic* means that motion is involved. The more mass that objects or athletes have, and in particular the faster they move, the more kinetic energy is produced.
- The greater the height and mass are, the more potential energy an object or athlete will have (height above the surface of the earth = distance, mass = resistance, and the force that accelerates them = gravity).
- Unlike objects that have kinetic energy, objects or athletes can have gravitational potential energy just by being positioned at a distance above the earth's surface.
- Gravitational potential energy is a form of stored energy that results from the acceleration of gravity.
- Strain energy, like gravitational potential energy, is a form of stored energy.
- In a collision, momentum is conserved by the colliding objects, whereas their kinetic energy is not.
- The coefficient of restitution (COR) is a measure of the ability of an object like a ball to spring back to its original shape after being hit.
- The two forms of friction are static friction, which exists between the contacting surfaces of two resting objects and provides the resistive force opposing the initiation of motion, and dynamic friction, which can be subdivided into two types of friction: sliding friction, a resistive force that develops when two objects slide and rub against each other, and rolling friction, which produces a resistive force when objects roll.

KEY TERMS

conservation of linear momentum

dissipation of kinetic energy

elasticity

energy

friction

gravitational potential energy

horsepower

impact

impulse

kinetic energy

Law of Conservation of Energy

power

rebound

rolling friction

sliding friction

static friction

strain energy

work

REFERENCES

MacKenzie, S.J., R.J. Lavers, and B.B. Wallace. 2014. "A Biomechanical Comparison of the Vertical Jump, Power Clean, and Jump Squat." *Journal of Sports Sciences* 32 (16): 1576-85.

Morio, C.Y.-M., L. Sissler, and N. Guégue. 2015. "Static vs. Dynamic Friction Coefficients, Which One to Use in Sports Footwear Research?" *Footwear Science* 7 (Suppl 1): S63-64.

Mytton, G., D. Archer, K. Thompson, and A.S.C. Gibson. 2013. "Reliability of Retrospective Performance Data in Elite 400-M Swimming and 1500-M Running." *British Journal of Sports Medicine* 47 (17): e4.

Smith, A., J. Roberts, E. Wallace, P.W. Kong, and S. Forrester. 2015. "Golf Coaches' Perceptions of Key Technical Swing Parameters Compared to Biomechanical Literature." *International Journal of Sports Science and Coaching* 10 (4): 739-55.

Stefani, R. 2014. "The Power-to-Weight Relationships and Efficiency Improvements of Olympic Champions in Athletics, Swimming and Rowing." *International Journal of Sports Science and Coaching* 9 (2): 271-86.

Tao, S. 2013. "Sports Equipment Based on High-Tech Materials." *Applied Mechanics and Materials* 340:378-81.

www Visit the web resource for review questions and practical activities for the chapter.

10

Moving Through Fluids

When you finish reading this chapter, you should be able to explain

- how the fluid forces of hydrostatic pressure, buoyancy, drag, and lift affect athletes and objects as they move through air and water;
- how pressure, temperature, and the nature of air and water affect the way that fluids act;
- how surface drag, form drag, and wave drag affect the movements of athletes and objects through air and water;
- what hydrostatic pressure and drag mean in sport;
- how competitors in high-velocity sports counteract or make use of drag and lift; and
- how the Magnus effect and drag forces affect the flight path of a spinning ball.

This chapter is about how the forces of air and water help or hinder an athlete. In the following pages you'll read how swimmers apply sport mechanics to use the resistance of water to maximize their propulsion, and you'll understand why one athlete can be a floater and another a sinker. You'll learn why competitive cyclists use aerobars, disc wheels, and slick racing suits and why discus throwers love to compete in stadiums where they can launch the discus into a headwind. You'll also read how a pitcher uses spin to produce a curveball and why golf balls travel farther and faster covered in dimples than they would if they had a smooth surface. This chapter is about applying mechanical principles to movement through the common fluids of air and water.

Hundreds of sports are affected by the forces produced by air and water. They range from the underwater sport of scuba diving to speed events like downhill skiing and auto racing, and from land-based events to high-flying sports like hot air ballooning and skydiving. In many sports, however, the forces exerted by air and water play little part in the outcome of the competition. Wrestlers don't worry about air resistance when they're pinning an opponent, and gymnasts don't move along a beam at speeds that require them to be streamlined. The same can be said about basketball. Basketball players would have to play on an outdoor court in a gale before they'd worry about the effect of wind on the quality of their performance.

Fundamental Principles of Moving Through Fluids

Generally, the forces that move through the fluids of air or water tend to flow, meaning that they will keep traveling in their initial direction. When an obstacle (like an athlete or an implement) is encountered, the air or water flows, or moves, around this obstacle. This flow can generate a range of forces; some will drag or slow down the movement, whereas others will create lift, buoyancy, or spin. The term *flow* conjures up an image of peaceful and tranquil movement, but the amount of flow is directly related to how fast the object is moving through the fluid.

To understand how these flowing forces work, look at figure 10.1, in which the flow lines around a ball are identified. This image will help you conceptualize an object's movement through a fluid, either air or water. In sport, we can envision this concept in cycling.

- If athletes are traveling at relatively slow velocity, say 8 km/h (5 mph), as they do during the recovery stage when riding a bicycle, they may sit upright to stretch their back and legs and grab a drink because the influence of airflow is minimal.
- When they return to sprint racing at six times the recovery velocity, at 48 km/h (30 mph), the influence of airflow plays an

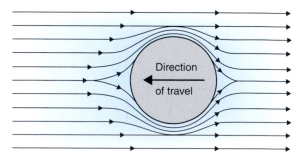

FIGURE 10.1 Fluid flow lines around a ball.

important role and they automatically drop down to the streamline position, tucked in tight behind the handlebars to "go with the flow."

As you can see from this example, how fast the athlete is traveling is a dominant factor when moving through a fluid. The other key factor is the type of fluid. Because water is almost 800 times denser than air, going with the flow in water is far more important than when traveling in air. To explain the key differences between these two fluids, we have separated them in the sections that follow.

Although air is obviously not as thick and dense as water, both act like fluids. They flow around and easily mold to the shape of an athlete or an object like the ball in figure 10.1 or the previous example of the cyclist. Air and water exert similar forces. These forces are hydrostatic pressure, buoyancy, drag, and lift. Let's look at each of these forces to see how they affect athletic performance.

Hydrostatic Pressure and Buoyancy

A fundamental force during movement through a fluid (air or water) is **hydrostatic pressure**, which is the force exerted by a fluid when supporting its own weight. This hydrostatic pressure is directly related to **buoyancy**.

- To be buoyant means to generate uplifting force, or, using the traditional definition, it is the "upward force on an object produced by the surrounding liquid or gas in which it is fully or partially immersed."
- This force is due to the pressure difference of the fluid between the top and bottom of the object.
- The net upward buoyancy force is equal to the magnitude of the mass of fluid displaced by the body.
- This force enables the object to float, or at least to seem lighter, as discussed in more detail in the following sections.

Hydrostatic Pressure in Air

To understand the characteristics of hydrostatic pressure, let's first look at the fluid air and our atmosphere. Most of us don't think of the atmosphere around us as pressing down on our bodies, but it does. In fact, gravity pulls the earth's atmosphere onto the surface of the earth, and because we move on the surface of the earth, the atmosphere presses on us, too.

An easy way to picture this concept is to think of the atmosphere as blankets layered on you when you're lying in bed. Each blanket holds up those above and presses down on those below. Now imagine that the surface of the earth is a bed with you resting on it. At sea level, you have the most blankets of atmosphere on top of you; at high altitudes, perhaps only one or two blankets are lying on you. In other words, the greatest atmospheric pressure exists at sea level, and it lessens as you increase altitude.

Most of our sporting applications are conducted close to sea level, so the influence of atmospheric pressure is minimal. At sea level, the weight of the atmosphere on the earth's surface is just over 100 kPa (kilopascals), or 14.7 psi (pounds on every square inch). That is 100 kilopascals pressing on every square centimeter of you, too!

If you measured the area of the top of your head, you could work out the weight of the atmosphere pressing down on your head. For adults, this weight can be more than 454 kg (1,000 lb). As humans, we've evolved so that we can tolerate this kind of pressure. You don't feel it, but we'd all be in big trouble if this pressure was ever removed.

- At sea level, the air we breathe is 21% oxygen and 78% nitrogen; the remaining 1% is made up of other gases.
- If you go from sea level to higher altitudes and compare the constituents of a **volume** of air, the air keeps to its ratio of 21% oxygen and 78% nitrogen.
- But air contains less oxygen and less nitrogen at higher altitudes because the air is at lower pressure; less oxygen and less nitrogen are squeezed into any given volume.

Climbers on Mt. Everest find that as they pass 7,925 m (26,000 ft) on their way to the peak at 8,848 m (29,028 ft), a third less oxygen is available in the air compared with what exists in a similar volume of air at sea level. For that reason, most Everest climbers carry additional supplies of bottled oxygen. Even at lower altitudes around 1,524 m (5,000 ft), becoming acclimatized to air that is at a lower pressure and gives you less oxygen each time you inhale is important. Tourists who casually decide to climb Mt. Kilimanjaro 5,895 m (19,340 ft) as part of a trip through Tanzania will find that they breathe more rapidly as their bodies attempt to get the oxygen they need. They frequently suffer from altitude sickness as they climb to the summit (Swann et al. 2016).

Hydrostatic Pressure in Water

At sea level water is almost 800 times denser than air, 784 times to be exact. Because water is virtually incompressible, the pressure it exerts increases with depth. Scuba divers (using their self-contained underwater breathing apparatus) learn that for every 10 m (33 ft) they go down in the ocean, the pressure exerted by seawater increases by the equivalent of one atmosphere (i.e., 100 kPa).

- So at 10 m you have 100 kPa, which is the weight of the atmosphere resting on the ocean, plus another 100 kPa as a result of the pressure exerted by the water. The total is two atmospheres, or 200 kPa (29.4 psi).
- At a depth of 20 m (66 ft), pressure increases to 300 kPa (44.1 psi), or three atmospheres.
- At 30 m (99 ft) the pressure is 400 kPa (58.8 psi), or four atmospheres, on every square centimeter of a diver's body (see figure 10.2).

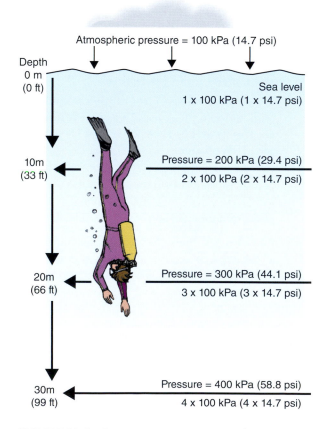

FIGURE 10.2 Pressure in water increases by one atmosphere (100 kPa) for every 10 m of depth.

Most of a diver's body can withstand this pressure, but not the air cavities. The sinuses, lungs, and inner ear all have air cavities, and pressure in these areas must be balanced with whatever pressure is exerted by the surrounding water. The type of water (fresh or salt) also influences the amount of pressure developed. Salt water has the added salt mineral, so this type of water is denser. A cubic meter (1 m × 1 m × 1 m) with its additional mineral content weighs approximately 1,025 kg (2,260 lb), which is 25 kg (55 lb) more than a cubic meter of freshwater. In imperial, a cubic foot (1 ft × 1 ft × 1 ft) of salt water weighs approximately 29 kg (64 lb), which is approximately 0.7 kg (1.6 lb) more than a cubic foot of freshwater.

When scuba divers descend in the ocean, they take with them compressed air in tanks, which passes through regulators designed to provide a balance in pressure between the air in the divers' lungs and air cavities and the pressure exerted by the water on their bodies.

- Air that is inhaled through the regulator at a depth of 10 m (33 ft) is reduced by the pressure existing at that depth to half the volume it would have on the surface of the ocean.
- At a depth of 20 m (66 ft), the air is reduced to one-third of its surface volume.
- At a depth of 30 m (99 ft), the air is reduced to one-quarter of its surface volume.

During ascent to the surface, the reverse occurs, and the compressed air expands as water pressure decreases.

- One lungful of compressed air inhaled at a depth of 10 m, if it's held in the lungs, will expand to twice its size at the surface.
- At 20 m, one lungful of air becomes three lungfuls at the surface.
- At 30 m, a single lungful expands to four times its size at the surface.

If a scuba diver fails to check his gauges before the dive, empties his tank at 30 m, panics, and decides to hold his breath all the way to the surface, he can expect the volume of air in his lungs to expand four times by the time he gets to the surface. The lungs cannot possibly handle this expansion. The unfortunate diver who forgets the cardinal rule of diving, "Never hold your breath," is likely to suffer an embolism in which expanding air ruptures the lungs and enters the bloodstream. Scuba divers avoid this dangerous situation by going with the flow and breathing normally on the way to the surface and by always having sufficient air to avoid rushed ascents.

The increased pressure that a diver experiences with depth also means that a tank of air that lasts an hour on the surface will last approximately 30 min at 10 m (33 ft), 20 min at 20 m (66 ft), and only 15 min at 30 m (99 ft). Not only is the diver's "bottom time" reduced the deeper he goes, but breathing air at such high pressures also allows high concentrations of nitrogen to get dissolved into the bloodstream, where it can affect the diver's central nervous system. Jacques-Yves Cousteau referred to the effect of nitrogen on a diver as "the rapture of the deep." The condition is also known as nitrogen narcosis, a state in which high concentrations of nitrogen make the diver act as though he is inebriated (Hemelryck et al. 2014). Other symptoms include loss of memory, slowed response time, faulty judgment, and disorientation.

- Nitrogen concentrations can also cause problems for a scuba diver on the way to the surface. If the diver rises too rapidly, the nitrogen that was dissolved in his blood under the higher pressure in deep water changes from a liquid state into tiny bubbles in the same way that carbonated water in a bottle fizzes when the cap is taken off.

- These nitrogen gas bubbles must be expired slowly through the lungs. But in the rush to the surface, the diver doesn't have enough time to expire the gas through the lungs. Instead, the bubbles can become lodged in the joints, muscles, and other tissues of the body.

- The result is a condition descriptively known as "the bends" because on the surface the diver can be crippled and bent over.

- Other symptoms of the bends, or more accurately decompression sickness, include disorientation, rashes, itching, dizziness, nausea, and pain in the joints.

- Divers normally avoid decompression sickness by having enough air in reserve to ascend slowly and by making timed decompression stops at various levels during the ascent, thereby providing sufficient time for the nitrogen to be safely transported to the lungs for elimination.

Scuba diving is much safer than it used to be. Advances in gas mixtures and equipment design and the availability of excellent training combine to make scuba diving a highly enjoyable sport. Two of the golden rules for scuba diving are

---

Free Diving

Free diving competitions are extreme sports in which athletes compete to achieve the greatest time, depth, or horizontal distance on a single breath without using scuba equipment. Among the six categories, the greatest publicity has been given to a category called No Limits. William Trubridge from New Zealand is the current unassisted free-diving world-record holder. He reached a depth of 100 m (328 ft) without assistance. By comparison, sperm whales, the world's deepest-diving mammals, have been recorded at depths in excess of 2,134 m (7,000 ft). Note that free diving is a dangerous activity that should be conducted only with appropriate supervision. As the name suggests, free diving is done without oxygen tanks; divers simply hold their breath underwater. Some of the injuries that can occur from the increase in pressure from diving are perforating or rupturing the eardrums or rupturing the eyeballs. Divers safely reduce the risk of this happening by equalizing the pressure in their sinuses and ears.

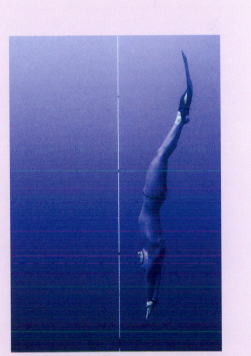

BORIS HORVAT/AFP/Getty Images

1. always dive with a partner and
2. understand the basic laws of physics as they apply to diving.

By following these rules, the scuba enthusiast can have many years of safe, pleasurable diving.

Buoyancy in Air

Although more common in a liquid, as discussed in the next section, buoyancy exists in the air. For example, hot-air ballooning is based on the fact that a volume of hot air weighs less than an equivalent amount of surrounding cooler air.

When you sit in the stadium seats and watch a television blimp circling overhead, you are looking at an object filled with a gas (helium) that is lighter than air. The blimp rises in the air because a buoyant, uplifting force is acting on it. It remains circling the stadium at a set altitude when buoyancy and other upward-lifting forces acting on the blimp are balanced against the downward pull of gravity.

Buoyancy in Water

In a liquid such as water, buoyancy is an extension of the concept of hydrostatic pressure. Because of the density of water, an uplifting force is generated, opposite to gravity, that pushes a submerged object upward. When an athlete (or an object) is fully immersed in water, the water presses on the athlete from all directions. Because pressure in a fluid increases with depth, more pressure is pushing on the athlete from below than from the sides or from above. The result is a **buoyant force**, a more powerful upward force pushing on the athlete from below.

Because numerous sports like swimming, rowing, kayaking, and yachting are affected by the buoyant force of water, let's look at how it works. A simple law known as **Archimedes' principle** will tell you how strong the upward lift of buoyancy is going to be. The buoyant force acting on an athlete is equal to the weight of the water that the athlete's body takes the place of, or pushes out of the way, when the athlete is immersed in the water (Pelz and Vergé 2014). When this buoyant force is greater than the force of gravity pulling the athlete down, the athlete is pushed to the surface (see figure 10.3).

Buoyancy will cause an athlete to float if he weighs 90 kg (200 lb) and the water he displaces weighs more than 90 kg. Because bone and muscle are more dense, their mass is more than the same volume of body fat, and an athlete with big bones, big muscles, and very little fat has a lot of mass compressed into the space that his body occupies. Therefore, an athlete with a body type that is predominantly bone and muscle is likely to have a greater mass than the water he displaces.

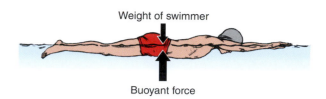

Weight of swimmer

Buoyant force

FIGURE 10.3 When the buoyant force is greater than gravity's downward pull (i.e., the athlete's mass), the athlete will float in water.

Consequently, gravity pulls down more than the buoyant force pushes up—and the athlete sinks. These types of athletes were once featured in televised superstar competitions. Huge, muscular athletes who were best at land sports turned the swimming competition into a submarine race because most of them were sinkers. The buoyant force acting on their bodies lost out to gravity.

But if these superstars had left the freshwater of the swimming pool and competed in the ocean, they would have been more likely to be floaters. Because of its additional salt content, as mentioned previously, the water that athletes displace in the ocean weighs more than the freshwater they displaced in the pool. If the amount of displaced ocean water weighs more than they do, they'll float.

In addition, the buoyant force acting on an athlete varies according to the density and temperature of the water. The warmer the water is, the less dense it becomes. Therefore, floating in a cold, salty ocean is easier than floating in warm freshwater. Cold ocean water is denser and has a greater mass than an equal volume of freshwater. In cold ocean water the amount of water displaced by an athlete is more likely to weigh more than the athlete does.

The percentage of body fat that athletes possess relative to bone and muscle plays an important role in distance swimming, particularly in cold ocean waters. Body fat not only assists in buoyancy but also keeps the athletes warm. Coupled with a grease coating that swimmers spread over their bodies, body fat helps resist the loss of body heat and the onset of hypothermia. Athletes with minimal body fat experience hypothermia more quickly than those who have a high proportion of body fat.

An athlete with a low percentage of body fat who puts on a wet suit will have an easier time staying afloat. A wet suit increases the athlete's space in the water with little increase in weight (because the wetsuit weighs very little). The thicker the wet suit is, the greater the space the athlete takes up in the water (i.e., the greater the athlete's volume) and the more water she displaces. The buoyant force pushing up increases, and the athlete's body mass doesn't change much. If an athlete wearing a wet suit descends well below the surface of the ocean, the increased pressure with depth tends to compress the suit. The deeper the athlete goes, the more compression occurs. In this situation the athlete can become a sinker.

Scuba divers wear lead hip weights to counter the additional buoyancy of the wet suit when they're close to the surface. As they descend, the increased pressure compresses the wet suit. Scuba divers must then balance the loss of buoyancy that occurs with increasing depth by pumping air from their tanks into what is called a buoyancy vest. The expanded buoyancy vest increases the space they take up in the water, so the buoyant force increases as well. Too much air in the buoyancy vest will cause the scuba divers to rise to the surface.

On the other hand, if they want to remain at a particular depth in the water, they pump just enough air into the vest to balance the upward push of the buoyant force against the downward pull of gravity. Anyone can use the lungs as a mini buoyancy vest.

- By taking a deep breath and expanding the rib cage, a person takes up more space in the water.
- The buoyant force increases without any change in body mass.
- For many athletes (but not all), this action is sufficient to hold them at the surface.

- Exhaling reverses the situation, and the person starts to sink. This phenomenon can be tested easily in a swimming pool.

The place where the buoyant force concentrates its upward push on an athlete's body or on any object (such as a boat) is called the **center of buoyancy**. An athlete's torso and upper body contain the lungs and collectively take up plenty of space in the water. Compared with the athlete's legs, the upper body takes up more space and generally has less mass than the water it displaces.

Consequently, the upper body is pushed upward more than the legs. The result is that the center of buoyancy for most athletes is not the same place as the center of gravity but a little farther toward the upper body, generally positioned just below the rib cage. If you could somehow hold the water in the manner and shape in which it is displaced by an athlete, you would find that the athlete's center of buoyancy is the center of gravity of the shape of the displaced water.

The position an athlete assumes when floating is determined by the fact that the center of buoyancy is usually closer to the upper body than the athlete's center of gravity is. Normally the buoyant force concentrates its upward push just below the rib cage, whereas gravity pulls downward at approximately waist level. These two forces, one acting up and the other down, cause rotation. The athlete's legs drop downward while the chest lifts up. Rotation ceases when the center of gravity is positioned directly below the center of buoyancy. The chest is up, and the lower body and legs hang beneath. Figure 10.4*a* shows gravity pulling down and buoyancy pushing up on a swimmer. The alignment of the pull of gravity and the buoyant force acting on the swimmer is shown in figure 10.4*b*.

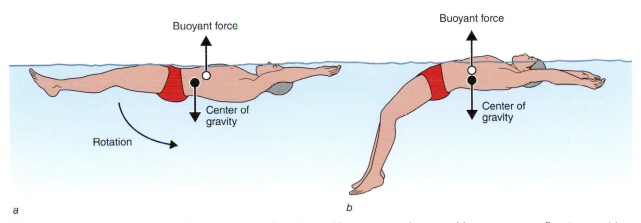

FIGURE 10.4   (*a*) Torque caused by the forces of gravity and buoyancy makes an athlete rotate to a floating position where (*b*) the center of buoyancy is directly above the center of gravity.

Adapted from K. Luttgens and K. Wells, *Kinesiology: Scientific Basis of Human Motion*, 8th ed. (New York: Times Mirror Higher Education Group, 1992), 364.

The tendency of an athlete's body to rotate when immersed in water is magnified when the athlete's legs are muscular and lean. A swimmer who has muscular legs with little fat must counter the legs' tendency to sink by lowering the head in the water as a counterweight and using an efficient leg kick to maintain a streamlined body position. Figure 10.5*a* shows an inefficient swimming position in which the athlete's body is angled downward with the legs low in the water. When the athlete's body is parallel to the direction of movement, the resistance from the water is reduced, as is the energy expenditure for propulsion (see figure 10.5*b*).

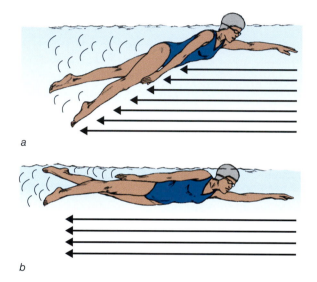

FIGURE 10.5   (*a*) In swimming, considerable drag results from a low leg position; (*b*) a high leg position reduces drag.

To maintain the stability of a boat, its center of gravity must be positioned well below its center of buoyancy. The keel (the bottom structure) on a sailboat counteracts the rotating effect of the waves and the wind blowing on the sails. A kayaker who has more mass in the upper body than in the hips and legs can raise the center of gravity (of the kayak plus its occupant) to a position where it is close to the center of buoyancy, or worse, to where it is positioned above the center of buoyancy. In this situation little force is needed to set up the turning effect of a torque sufficient to flip the kayaker upside down. Hanging below the kayak and fully immersed in water, the kayaker now becomes the equivalent of the keel on a sailboat.

To spin the kayak 180° and raise himself out of the water, the immersed kayaker must flex his body upward toward the kayak so that his center of gravity is as close as possible to the center of buoyancy. After reaching this position, the kayaker uses the paddle and a sharp hip thrust to spin himself and the kayak so that he lifts himself out of the water. This action, commonly called an Eskimo roll, is much easier to perform in agile river kayaks than in larger touring ocean kayaks.

Drag and Lift

When we are moving through a fluid, the individual forces of drag and lift combine to produce a resultant force. The **drag** forces are in the direction of travel, and they oppose, or drag against, motion. The **lift** forces are perpendicular to the drag force, as shown in figure 10.6. Together, these two com-

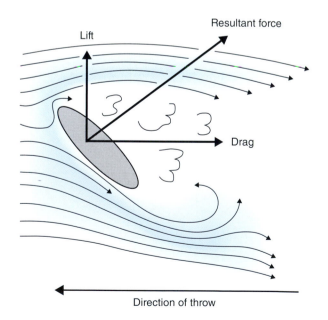

FIGURE 10.6   Lift and drag forces combine to form a resultant force.

bine to generate a resultant force. In the following sections we discuss the various drag and lift forces separately.

Drag and Relative Motion

In most cases, drag is "a drag" because it's a collection of fluid forces that tend to oppose the actions that an athlete is trying to perform. Drag pushes, pulls, and tugs on an athlete.

- If an athlete runs to the left, drag forces act to the right. Whatever the direction, drag forces act in opposition. It's easy to guess what will increase these drag forces.
- If athletes sprint faster, the air will push and pull at them more. If they sprint down the beach and then into the water, as occurs at the start of triathlons, the drag on their bodies is greater in the water than in the air.

Drag varies according to

- whether the fluid is water or air,
- whether the fluid is warm or cold, and
- how dense and viscous (sticky and clinging) it is.

Drag also depends on the size, shape, and surface texture of the athlete and likewise on the size, shape, and surfaces of the equipment that the athlete is using. A giant sprinter running in a full-length fur coat that grabs at the wind will generate considerably more drag than a smaller athlete who wears a slick, polished bodysuit.

When you consider drag, remember that what or who is doing the moving doesn't matter—it can be the athlete, the fluid, or both. If the fluid is moving, then drag occurs even when an athlete is standing still. What's important is the relative motion (i.e., how quickly the fluid and athlete are passing each other).

Let's look further at this concept of **relative motion**. Figure 10.7 shows three variations in relative motion. In figure 10.7*a*, an athlete is sprinting at 32 km/h (20 mph) into a headwind of 8 km/h (5 mph). The athlete moves in one direction at 32 km/h, and airflow is in the opposite direction at 8 km/h. The relative velocity of airflow past the athlete in this situation is 40 km/h (25 mph). In figure 10.7*b*, the athlete sprints at 32 km/h but with a tailwind of 8 km/h. The relative velocity of airflow past the athlete is now 24 km/h (15 mph). In figure 10.7*c*, the athlete is again sprinting at 32 km/h but with a tailwind of 32 km/h. The relative velocity of airflow past the athlete in this situation is zero, because the athlete and airflow are traveling in the same direction at the same velocity. In the three examples we've just described, the frictional forces acting on the athlete occur maximally in figure 10.7*a* and minimally in figure 10.7*c*.

Drag and Relative Motion in Air

Wind tunnels are used to assess drag forces that occur when objects move through air. Wind tunnels allow the aerodynamic qualities of planes, race cars, speed skaters, ski jumpers, cyclists, or any object affected by the forces exerted by airflow to be observed. For example, when air is driven past sprint cyclists in a wind tunnel, assessments can be made of the drag forces that occur with different bike designs, various types of competitive

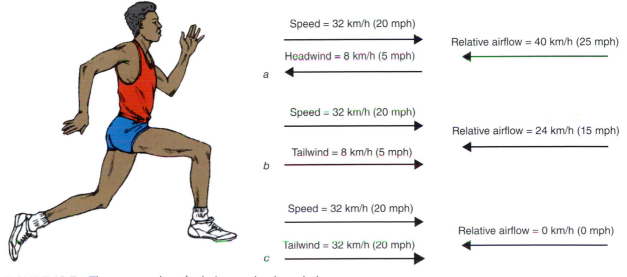

Speed = 32 km/h (20 mph)
Relative airflow = 40 km/h (25 mph)
Headwind = 8 km/h (5 mph)
a

Speed = 32 km/h (20 mph)
Relative airflow = 24 km/h (15 mph)
Tailwind = 8 km/h (5 mph)
b

Speed = 32 km/h (20 mph)
Relative airflow = 0 km/h (0 mph)
Tailwind = 32 km/h (20 mph)
c

FIGURE 10.7   Three examples of relative motion in sprinting.

clothing, and different body positions used by the cyclist.

Wind tunnels give valuable data on how to reduce the drag forces generated by airflow over a particular object, along with information on the shapes and body positions that provide optimal flight characteristics. This information is then used in such sports as gliding and ski jumping and in throwing events like javelin and discus.

Most discus throwers know that a discus travels farther when it is spun at release and launched (like a glider) at an appropriate angle into a headwind. Wind tunnel tests can determine the optimal trajectory and spin of the discus relative to wind velocity and wind direction (Asai et al. 2007). From wind tunnel tests using a spinning discus and from practical experimentation by throwers, it has been found that a headwind of 24 to 32 km/h (15 to 20 mph) can add 6 m (20 ft) or more to the distance thrown, as compared with throwing in still air.

Consequently, athletes try to adjust the flight of the discus according to the conditions they meet. Discus throwers quickly learn which venues have winds that blow regularly from a favorable direction. They then try to compete at such venues as often as possible. Currently, no wind speeds are illegal in relation to setting records in discus, javelin, hammer, and shot put, in contrast to wind-speed standards set for sprint races and long jump.

Drag and Relative Motion in Water

High-performance swimmers frequently practice in special observation swim tanks called flumes. The coach views their strokes through a side window or underwater camera. In a flume, a turbine drives water past the swimmer's body. The speed of the turbine can be adjusted to drive the water at various velocities toward the swimmer. Because the water is moving, the athlete must adjust her swimming rate to stay in the same spot.

In a swimming pool the reverse occurs. The water is stationary, and swimmers travel through and past the water. A flowing river presents a situation in which both the water and the swimmers move. These three situations (swimming in a flume, a pool, and a river) provide the three possible variations in what is called relative motion, which is the motion of one object relative to another. In these three situations, one element is a swimmer and the other is the water.

- If the relative movement of swimmer and water past each other is the same, then the drag forces produced by this movement represent this process.

- In other words, drag forces can be the same, no matter who or what is doing the moving.

Surface Drag in Air and Water

Surface drag is also called viscous drag and skin friction. These names immediately give us a clue to the characteristics of surface drag. When an athlete or object moves through a fluid such as air or water, the fluid forms what is called a **boundary layer**, the layer of air or water that, as it passes by, comes in direct contact with the athlete or the object, causing surface drag. Because of the fluid's viscosity, it clings and drags at the surface of whatever it contacts. Figure 10.8 shows the boundary layer around a ball that is moving slowly through the air.

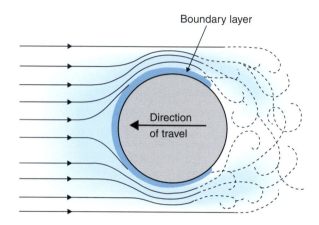

FIGURE 10.8 Boundary layer around a ball traveling through the air.

Viscosity is a measure of a fluid's flow. You can think of viscosity as a fluid's stickiness, its resistance to flow, and its ability to cling to the surface of any object that passes through it.

Air is obviously less sticky than water, but air differs from water in that its viscosity increases slightly with an increase in temperature, whereas the viscosity of water decreases with an increase in temperature.

- The viscosity of water and air plays a big part in what is called surface drag, and with an increase in viscosity, surface drag increases.

- Surface drag combines with two other types of drag, form drag and wave drag, which are discussed next. These three drag forces are the most important ones that affect the movement of an athlete or object.

The amount of surface drag developed by the movement of an athlete or an object through the air, such as a sprint cyclist pedaling a bike, depends on how viscous or sticky the air is and how large the surface area of the athlete and the bike is that comes in contact with the air flowing by. Surface drag also depends on the roughness of the surfaces of the athlete and the bike that contact the air. Finally, surface drag depends on the relative velocity of the cyclist and the air as they pass each other. An increase in any of these factors increases surface drag.

Form Drag

To surface drag, as discussed in the previous section, we now add its close associate, form drag, also called shape drag, profile drag, and pressure drag. As these names suggest, **form drag** is produced by the shape and size of an athlete or object, such as a race car as it passes through the air.

- An area of high pressure develops in front where the athlete or object pushes against the air head on.

- Immediately to the rear is an area of turbulence. This trailing area, called the **wake** (like the wake to the rear of a ship), is a region of swirling low pressure and suction.

- With high pressure at the front and low pressure at the rear, an imbalance in pressure exists from front to rear that opposes forward motion. This effect is called form drag or, to use the more descriptive term, pressure drag.

Form drag increases if the relative velocity of the athlete (or object) and the air as they meet and pass each other increases and if the shape of the athlete or object promotes development of high pressure in front and low pressure to the rear. What kinds of shapes are we talking about here?

Think of an athlete pedaling as fast as possible into a strong wind sitting bolt upright on an old-fashioned bicycle. An upright body moving at high velocity would cause a large high-pressure area to develop in front of the athlete and a huge turbulent low-pressure area to develop to the rear. If the athlete wore loose, flapping clothing and strapped huge carrier bags and luggage to his bike, these items would increase not only his form drag but also his surface drag.

Efforts to reduce form and surface drag as much as possible occur in many sports in which athletes and their equipment move at high velocity. In these sports you see body positions and equipment that are designed to

- eliminate a blunt shape with a large cross-sectional area that must push through the air at right angles to the direction of fluid flow;

- eliminate lumps, bumps, projections, and rough edges and instead smooth out and polish all surfaces that contact the flow of the air; and

- eliminate the turbulent wake that occurs at the rear of the athlete or object where the low pressure occurs.

Form Drag in Air

Designers working in high-velocity sports such as cycling, downhill skiing, auto racing, luge, and boat racing aim for a streamlined teardrop airfoil shape that also has ultrasmooth surfaces.

- Figure 10.9 *a* and 10.9*b* compare the flow patterns around the blunt circular shape of a ball or a cylinder with that of a teardrop shape.

- Notice in figure 10.9*a* that a teardrop shape causes the flow patterns to eliminate this turbulent wake.

- In figure 10.9*b* note that a circular shape produces a turbulent wake area to its rear.

- The teardrop shape could be improved further if it were made more dart shaped. Coupled with supersmooth surfaces, this shape would reduce drag even further.

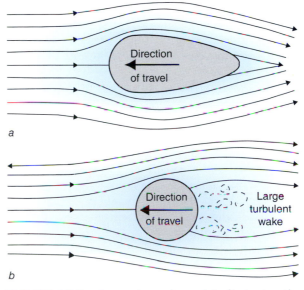

FIGURE 10.9 A teardrop shape (*a*) eliminates the turbulent low-pressure wake that occurs to the rear of a circular shape, such as a ball (*b*).

Let's look closely at the sport of competitive cycling to see how drag is reduced to minimum levels. Elite athletes shed excess body weight and ride bikes made of lightweight metals and modern composite materials to reduce nonproductive weight to a minimum. The bike is often raked (i.e., the frame is tilted downward toward the front wheel) and equipped with a larger rear wheel and extended handlebars. The combination of these features forces the rider into an aerodynamic position with the back parallel to the ground. The athlete's head is lowered, the arms are extended, and the hands are together in front. This flattened cycling position reduces the frontal area (and high pressure) to a minimum and helps the cyclist simulate an airfoil shape to cut through the air with the hands and arms leading the way.

To reduce drag generated by surface friction, the athlete wears a superslick, low-friction one-piece skintight suit with no wrinkles or loose flapping sections. Skintight fingerless gloves and laceless shoes, or booties, are also used. Legs, arms, and face are shaved. A teardrop aerodynamic helmet is worn. Frequently the helmet has a visor section in front to help slice through the air and a long blade shape to the rear to reduce turbulence and low pressure.

A disc rear wheel is used, which eliminates any drag caused by spokes. The front wheel is made of lightweight composite materials with three large spokes that are teardrop shaped in cross section. An alternative design for the front wheel uses a low number (14) of high-tension, blade-shaped spokes. These design features are intended to reduce the tremendous drag generated by the normal number (36) of circular spokes found on each wheel of a standard bicycle. All other components on an aerodynamic bike are teardrop shaped. Shapes that are circular in cross section, like the tubing on a standard bicycle, are avoided at all costs. Athletes looking for the ultimate in slipping through the air spray their bodies with silicone!

Even cyclists who use all these modifications to conquer ineffective positions and various types of drag are not as aerodynamically efficient as those who cycle in a recumbent position (i.e., inclined backward) and who cover themselves totally with lightweight fiberglass shells. Athletes enclosed in this manner have pedaled in excess of 128 km/h (80 mph), far faster than the 72 km/h (45 mph) reached by the world's best sprint cyclists.

Classified as HPVs (human-powered vehicles), lightweight shells are not allowed in the Tour de France or in Olympic cycling competitions. For the average person, cycling in a recumbent position inside a lightweight shell makes cycling faster and less energy consuming. But being lower down, these cyclists are difficult for motorists to see, and the vehicles are unstable in crosswinds.

Standing in front of one of these teardrop-shaped human-powered vehicles, you'll notice how narrow and long they are. Their narrow, pointed profile reduces the cross-sectional area at the front of the vehicle and the high pressure that occurs there. The sides of the shells and fairings are smooth, polished, and slick, and all vehicles are designed with a long, tapering tail. The tapering tail fills in the low-pressure wake that occurs to the rear of the vehicle. The objective is the same as that of the speed skier, to reduce form and surface drag to a minimum (see figure 10.10).

Similar to those who use human-powered vehicles, luge competitors lying back on their small sleds wear form-fitting, slick uniforms and even try to reduce the drag caused by the shape of their faces. Although the nose is small compared with the rest of the athlete's body, it's still a projection and increases drag. To counter this problem, athletes cover their faces with plastic shields. Wearing a shield is worth the effort when 0.01 s frequently separates one medal from the next.

Form Drag in Water

As mentioned previously, because of the greater density of water, the influence of form drag during movement through water is naturally of greater importance. Elite swimmers, divers, rowers, and kayakers each apply the same streamline-form shape as discussed for traveling through air and shown in figure 10.9a. They aim to produce the minimum amount of turbulence, particularly in the wake (Formosa et al. 2014). The shipbuilding industry allocates a tremendous amount of resources to reducing the form drag of seagoing vessels.

These efforts include testing the size and shape of the keel and bow of the vessel. For example, in the 2003 America's Cup yacht race, the beam (width of the yacht) for the top four vessels was 0.2 m (0.7 ft) narrower than their original design, thus reducing the form drag of the yacht. The most recent short and wide Oracle bulb keels are an extension of the radical wing keel unveiled back in the 1983 America's Cup race. The slope, shape, and size of the keels for the Swiss, United States, and New Zealand boats in the 2003 regatta were also all different, highlighting the influence of form drag on going with the flow. In the 2017 America's Cup, the New Zealand and American teams raced

FIGURE 10.10 Aerodynamically designed high-speed human-powered vehicle. The teardrop shape reduces aerodynamic drag.

CATRINUS VAN DER VEEN/AFP/Getty Images

in the final with radically designed multihull catamaran hulls. These boats had foil sails, negative stability to counter any movement on the boat, and aerospace construction to reduce weight and increase strength.

AT A GLANCE
Drag and Lift Forces

- When moving through a fluid, the individual forces of drag and lift combine to produce a resultant force. The drag forces are in the direction of travel, and they oppose, or drag against, motion. The lift forces are perpendicular to the drag force.

- Athletes can manipulate their flight path (or the implement they are using) to take advantage of the drag and lift forces.

- The three types of drag are surface drag (also called viscous drag and skin friction), form drag (also called shape drag, profile drag, or pressure drag), and wave drag.

Wave Drag in Air and Water

The third and final type of drag acts at the interface where water and air meet. This type of drag is called **wave drag**, and it affects swimming (particularly breaststroke and butterfly, which can involve a lot of up-and-down motion) and other water sports like rowing, kayaking, speedboat racing, and yachting. Because the drag occurs at the air–water interface, delineating two categories of wave drag is not possible.

When swimmers move through the water, waves pile up in front of them and create a high-pressure wall of water that resists their forward motion. The faster they travel and the larger the cross-sectional area that they push forward against the water, the larger the wall of water that builds up in opposition. So if a swimmer acts like a barge bashing its way through the water, the wave drag will be phenomenal.

- At competitive speeds, wave drag is much more detrimental and energy consuming than surface and form drag.

- Surface and form drag increase according to the square of the velocity. This factor is important to consider.

- Therefore, if a swimmer doubles her velocity through the water, both surface drag and form drag increase fourfold.

- Wave drag, on the other hand, increases according to the cube of the velocity. If a swimmer doubles her velocity, then whatever

Transferring Forces in a Boat

To create velocity within a boat, any forces generated (such as the wind on the sails) must be transferred down the structure of the boat with minimal loss. Any force that is not translated in the boat will be dispersed away from the boat and may not contribute to boat speed. An example of the importance of this transfer was seen in new America's Cup rigid foil sail design. This solid sail increased the efficiency of the boat by more effectively transferring the forces of the wind into boat speed. Adjusting and setting the sails on these large boats require the use of winches. A typical boat winch moves in the horizontal plane and is driven with one hand. To increase the transfer of forces, racing yachts introduced "grinders." These hand cranks (much like a bicycle pedal setup) moved in the vertical plane and allowed two hands to wind the sail in and out. At the 2017 America's Cup the winning New Zealand boat increased this force transfer further by using cycling ergometers to wind the sails in and out. Because an athlete's legs are more powerful and can generate more forces than the arms, cycling, when compared with hand cranking, can generate more forces within the boat.

Josh Edelson/AFP/Getty Images

wave drag she was generating at the original velocity increases eightfold (i.e., $2 \times 2 \times 2$).

- If the swimmer triples her velocity, the original amount of wave drag increases 27 times (i.e., $3 \times 3 \times 3$).

Inefficient swimming technique and poor swimming pool designs produce waves. A top-rated swimming pool has specially designed gutters and lane dividers that absorb waves and stop them from bouncing and flowing from one lane to the next. This criterion is one of those used to classify a competition swimming pool as a fast pool because it reduces wave drag and thus naturally allows faster swimming times.

When an athlete's swimming technique is poor, wave drag is likely to increase along with other forms of drag. In events like the famous Hawaiian Ironman and Iron Woman triathlons, wave drag caused by the ocean and by other competitors can be a significant form of resistance. Expert swimmers hone their swimming technique to cut wave drag to a minimum. They also learn to draft in the wake of a leading swimmer. This tactic simulates drafting used by cyclists and allows them to swim with less expenditure of energy. But they risk getting kicked in the face! There are two ways of beating wave drag.

- One is for the athlete to swim underwater as long as possible. In the past this technique was used in breaststroke, backstroke, and butterfly; swimmers would stay underwater for almost the complete lap. To get the swimmer back on top of the water, the current rules limit the underwater distance to 15 m.

- The second method of beating wave drag is to get as far out of the water as possible (i.e., to hydroplane). This technique is used by high-speed boats; the pressure that the water exerts on the planing surfaces of the boat lifts the hull partially or completely out of the water. This position reduces the wave, surface, and form drag caused by the water. The boat fights against the lesser resistance generated by moving through the air.

How Drag Affects the Flight of Baseballs, Tennis Balls, and Golf Balls

If you examine what happens to a smooth-surfaced ball as it travels through the air and what happens to a ball with seams (like a baseball) or dimples (like a golf ball), you'll find some dramatic differences in the way that drag affects the balls. These variations play an important role in determining the flight of baseballs, tennis balls, and golf balls (Jinji and Sakurai 2006).

A circular, blunt object like a ball is poorly streamlined compared with a teardrop-shaped object; in fact, all circular shapes produce considerable drag, as shown in figure 10.11*a* and 10.11*b*. Furthermore, you can't change the shape of a ball the way that you can redesign a race car, a bobsled, or a bike with cross-sectional tubing to make them more aerodynamic. A ball is always ball shaped, no matter how small or large.

To understand how drag forces affect the flight of balls, let's have a smooth-surfaced ball move through the air very slowly. At this velocity the boundary layer of air that contacts the surface of the ball flows smoothly around the ball. Its flow pattern looks like laminations in a piece of plywood. This type of pattern is called **laminar flow**, or streamlined flow, illustrated previously in figure 10.1. Moving at low velocity, the ball is

affected predominantly by surface drag caused by the clinging, viscous nature of the boundary layer.

If we make the ball travel faster so that the velocity of airflow around it increases, the laminar flow starts to break up. Smooth laminar flow is now mixed with a disturbed, distorted **turbulent flow** pattern. Because the air passes the ball at higher velocity, the air cannot follow the ball's contours in the same way it did when it passed slowly. Instead, the boundary layer follows the ball's contours only partway; thereafter it tears away from the ball's surface toward the rear of the ball. A turbulent low-pressure wake develops at the rear, as shown in figure 10.11*a*.

- With the increase in the velocity at which air and ball pass each other, pressure also builds up where the ball hits the air head on.
- So we have high pressure increasing at the front and low pressure increasing at the rear. The result is that the ball experiences an increase in form drag.
- If the ball and the air travel by each other at even greater velocity, the place where the boundary layer breaks away from the ball's surface moves from the rear of the ball toward the front.
- The result is an even bigger wake area at the rear, as shown in figure 10.11*b*. More high pressure at the front of the ball and more low pressure at the rear cause form drag to increase even further.

Finally, if the ball and air pass each other at extremely high velocity, the boundary layer becomes extremely turbulent. Now a surprising change happens. When the boundary layer is extremely turbulent, as it is around the dimples on a golf ball, for example, the place where it separates from the ball shifts back toward the rear and in so doing reduces the size of the low-pressure wake, as shown in figure 10.12.

Because the low-pressure wake has been reduced, the ball's form drag is reduced as well, even though the ball's surface drag will have increased. These variations in form and surface drag are important in games like baseball and golf because seams or dimples cause a turbulent boundary layer to form all around the ball—not only at high velocities but also at low velocities.

Therefore, seams and dimples reduce form drag and help the ball fly farther and faster and hold to its flight path better than if it were smooth all over.

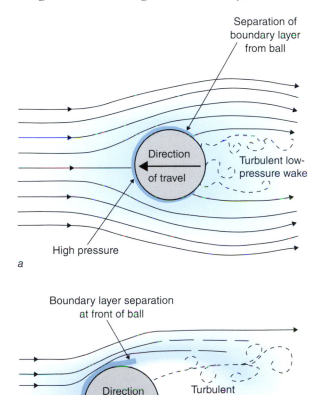

FIGURE 10.11 Turbulent flow pattern. (a) At high velocities, the boundary layer breaks away to the rear of the ball, and (b) at even higher velocities, the boundary layer shifts to the front of the ball.

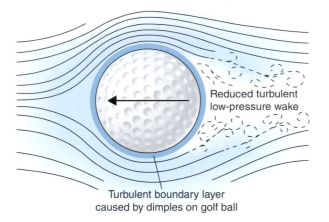

Reduced turbulent
low-pressure wake

Turbulent boundary layer
caused by dimples on golf ball

FIGURE 10.12 Turbulent flow pattern. At extremely high velocity, the boundary layer becomes fully turbulent, reducing the size of the low-pressure wake. As a result, form drag decreases.

- Although roughing up the surface of a ball that has seams and dimples causes surface drag to increase, the total drag affecting the ball is less because the bigger drag (i.e., form drag) has been reduced.

- When spin is put on the ball (an occurrence that is a big part of baseball and golf), a ball with seams and surface roughness curves better in flight and holds to the curve better than a ball that is smooth.

Lift

To understand the theory behind the creation of lift as a propulsive force, let's see how it occurs when air flows over an airplane's wing. A cross section of a plane's wing is called an **airfoil**. Airfoils come in all shapes and sizes. Some are fat, some are thin, some are symmetrical, and many are built with an undersurface that is predominantly straight and an upper surface that is curved.

- To generate lift, an airfoil is tilted (i.e., angled) relative to the flow of the air passing over it.

- This angle of tilt is called the **angle of attack**. Angles of up to 15° are considered most advantageous for generating lift.

- Research has shown that air passing over the upper surface of an airfoil increases in velocity as it flows from the leading edge to the trailing edge of the airfoil.

- In addition, the airflow is deflected downward by the inclination of the upper surface of the airfoil. On the undersurface, the

velocity of the air is reduced and deflected downward as it follows the lower surface of the airfoil.

Thanks to the work of Daniel Bernoulli, a Swiss mathematician, we know that a decrease in the velocity of a fluid increases the pressure exerted by that fluid, and conversely that an increase in the velocity of a fluid decreases fluid pressure. This concept is known as **Bernoulli's principle**. Applied to the airfoil, air pressure below the airfoil is increased and air pressure above the airfoil is decreased (see figure 10.13). This pressure gradient, acting from below to above, helps produce the force of lift.

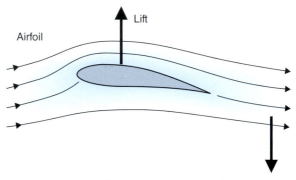

Lift

Airfoil

FIGURE 10.13 Lift generated by an airfoil (i.e., a cross section of an airplane's wing). The pressure above the airfoil is less than that below. Airflow both above and below the airfoil is directed downward, contributing to lift.

Lift is further enhanced by a reaction to the motion of the air as it is directed downward from both the upper and lower surfaces of the airfoil. If you direct air molecules downward, meaning that you increase the angle of attack, an equal and opposite reaction occurs in the opposing direction. Therefore, Bernoulli's principle and Newton's laws work together to apply a lift force to the airfoil (see figure 10.14), and the same process occurs in water.

Finally, the consistency of the fluid and its relative motion dramatically influence lift. When air and water are thicker (denser), lift increases. Also, the faster the fluid moves by an object, the greater the lift is. A water skier will therefore experience more lift (but also more drag) when skimming across the ocean compared with a freshwater lake. Similarly, a ski jumper will get more lift at lower altitudes, where the air is denser, than in the thinner air of higher altitudes.

Lift in Air

When an athlete throws a discus, the discus applies force to the air, and the air reacts by applying force

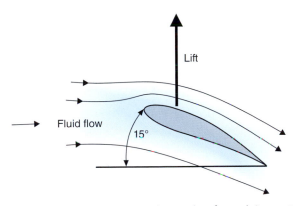

FIGURE 10.14 Increasing the angle of attack (i.e., raising the leading edge of the wing relative to the flow of air) up to 15° increases lift.

to the discus. The force that the air exerts can be broken down into two separate forces. One acts in the same direction as the airflow, directly counteracting the implement's forward movement. This effect is the drag force we discussed earlier. The other force acts at right angles to the drag force and is called lift.

Experimentation in a wind tunnel with the tilt of the discus relative to the airflow (called the angle of attack) indicates that variations in this angle determine how much lift and drag a discus undergoes. When the leading edge of the discus is tilted upward, as it is in figure 10.15, air is deflected downward and exerts an equal and opposite pressure upward. This equal and opposite pressure produces the upward lift acting on the discus.

The leading edge of a discus can be angled upward only so far before lift disappears. If the angle is too great, lift disappears and drag increases dramatically, causing the discus to stall.

- Figure 10.15a shows a worst-case scenario. A discus thrown in this fashion gets no lift at all and quickly drops to the ground.
- If the leading edge of the discus is tilted downward, as shown in figure 10.15b, the lift force acts downward. As strange as it may seem, this effect is still called lift.
- Lift does not have to be upward, although the term itself suggests this. Lift can occur in any direction.
- The resultant of lift and drag forces acts backward and downward, opposing the motion of the discus.

Many factors influence the amount of lift acting not only on sport implements like the discus but also on the athlete. Lift is influenced by the athlete's body position because shape influences airflow patterns. The body position assumed by the great Norwegian, Finnish, and Japanese ski jumpers traveling down the in-run is far different from the position they use immediately after takeoff. Accelerating down the in-run, they crouch low with their backs parallel to the in-run. This body position is similar to that used by competitive cyclists in that it minimizes air pressing against their chests and slowing them down. When they take off, they extend their bodies and angle them forward and

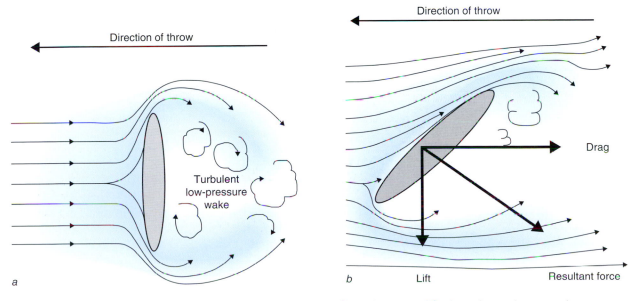

FIGURE 10.15 (a) A discus thrown with too great an angle of attack gets no lift. (b) A discus thrown with a negative angle of attack generates downward lift.

upward, giving an angle of attack that maximizes lift. By holding this position, they use lift to extend their time in the air, producing soaring jumps of well over 121 m (400 ft).

- The faster that a ski jumper flies through the air, as shown in figure 10.16, the greater the lift force is that pushes him upward.

- This force depends on the ski jumper's holding an optimal body angle to maximize lift. The jumper must change this angle as he slows in flight.

- Likewise, the size of the surface area that the jumper angles into the airflow influences lift. If this area is increased, the lifting force increases in much the same way that a larger wing surface on a plane increases lift.

- Ski jumpers now angle their skis in a V shape so that the skis and the jumper's body both contribute to lift. The old style of jumping with the skis parallel beneath the jumper's body is inefficient, and jumpers no longer use this technique.

The amount of lift that an athlete requires varies according to the demands of the sport. Downhill speed skiers compete to see who can reach the greatest speed, often faster than 209 km/h (130 mph), through a measured section of the course. Lift is the last thing these athletes want because it causes them to lose contact with the snow. Their skis are purposely long and heavy, and tips have hardly any curl so that they do not lift upward. Speed skiers fight to keep their upper bodies parallel to the ground so that the lift from their upper bodies is minimal. Their body positions and equipment design are intended to keep them locked onto the snow.

The action of the skis in water skiing differs from that in speed skiing because lift is a necessity. Unless the athlete gets lift she has no hope of rising onto the surface of the water. The boat must pull the athlete at an adequate velocity to help produce lift. The faster the boat goes, the more lift is produced. In addition, water skiers angle the tips of their wide water skis more at the start than when traveling on the surface. When they do this, the reaction force from the water combines with the pull of the boat to lift the athlete out of the water and into a skiing position (see figure 10.17).

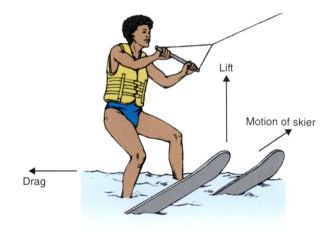

FIGURE 10.17 Lift and drag in water skiing.

Downward lift (as well as the resultant forces generated by drag and lift) is commonly made use of in auto racing, where rear spoilers are angled to press the car down onto the road. This effect improves the friction of the tires with the road surface and gives better traction (see figure 10.18).

Lift in Water

Earlier in this chapter we looked at how drag forces commonly oppose the motion of an athlete. In this section you'll see that it's possible to get drag and lift to act as propulsive forces. We'll see how this occurs in the sport of swimming.

In the late 1960s it was thought that swimmers should pull and push directly back against the water to propel themselves in the opposing direction. If an athlete is swimming to the left, the thrust from the swimmer's hands and arms to the right against the water generates a drag force (i.e., a reaction force) from the water that propels the swimmer forward. Figure 10.19 shows a lateral view of a swimmer's hand moving to the right and the drag force that propels the swimmer, acting to the left.

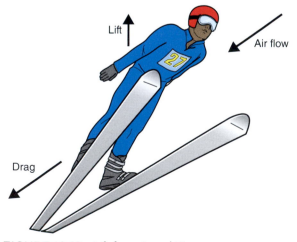

FIGURE 10.16 Lift force in a ski jump.

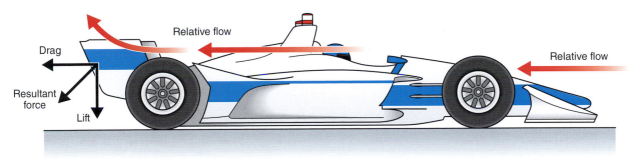

FIGURE 10.18   Lift, drag, and the resultant force produced by spoilers press a race car down onto the track and improve its traction.

Adapted from J.G. Hay and J.G. Reid, *Anatomy, Mechanics, and Human Motion*, 2nd ed. (Englewood Cliffs, NJ: Prentice-Hall, 1988), 228.

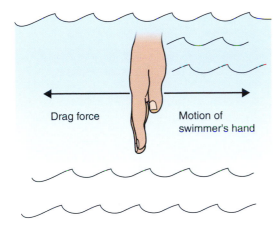

FIGURE 10.19   The motion of the swimmer's hand pushing the water to the right creates a drag force that propels the swimmer to the left.

In the 1970s and 1980s **propulsive drag** was challenged by those who thought that hydrodynamic lift provided the dominant propulsive thrust for a swimmer. It was proposed that rather than push directly backward to swim forward, swimmers should concentrate on performing a weaving down–in–up–out S-shaped pull and push motion with the hand and arm to produce lift.

Advocates of lift as a propulsive force in swimming noticed the tendency of swimmers to use a weaving S-shaped pull and push with the hand and arm. This action occurred in all strokes, particularly in freestyle and butterfly.

- It was thought that during these weaving down–in–up–out motions, the swimmer was using the hands as airfoils and that water flowing over the upper surfaces and undersurfaces of the hands produced a lift force in the same way that air does when flowing over and under an airfoil.

- By varying the angle of the hand during the pull and push of the stroke, as shown in figure 10.20, the swimmer achieved the optimal angle of attack in the same way as an airfoil moving through the air.

- These hand movements were described as "blading" motions because it was thought that swimmers were using their hands to generate lift the way that the airfoil-shaped blades of a boat's propeller do.

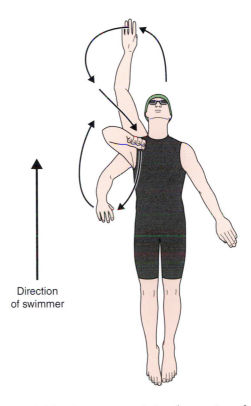

FIGURE 10.20   A swimmer mimics the motion of propeller blades to develop a propulsive lift force.

Drafting in Cycling

Nothing demonstrates the advantages of drafting more clearly than comparing the top speeds attained by cyclists with and without drafting. Drafting occurs when the cyclist rides close behind another rider and drafts in the other's slipstream, thereby requiring less force to propel the bicycle. When cycling in a velodrome with no help from a pace vehicle, the top velocity is up to 80 km/h (50 mph). For a cyclist pedaling in a recumbent position enclosed in an aerodynamic shell and riding at high altitude, the top speed is just over 128 km/h (80 mph). Compare these speeds with the incredible 269 km/h (166.9 mph) achieved by Dutchman Fred Rompelberg in 1995 when he drafted in the lower-pressure suction zone behind a specially designed race car on the Bonneville Salt Flats. This effort is the current cycling speed world record. Rompelberg's bike was equipped with two chain-wheels that stepped up the revolutions occurring at the back wheel. With this system, one revolution of the pedals moved the bike a phenomenal 32 m (35 yd). With such a gear system, Rompelberg could not accelerate the bike from a dead stop. Instead he had to be towed to 137 km/h (85 mph) and then released from the rear of the pace vehicle.

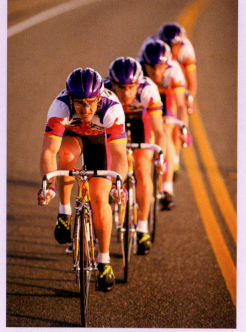

John P Kelly/The Image Bank/Getty Images

Modern research has shown that although lift acts as a propulsive force during certain phases of a swimmer's hand and arm motions, it does not contribute as much to propulsion as drag propulsion does. Lateral and vertical in–out–up–down hand actions still occur because of other motions the swimmer makes, such as arm recovery at the end of the stroke and rolling of the head and body for breathing. In strokes like freestyle, backstroke, and butterfly, coaches no longer emphasize excessive lateral and vertical hand movements with the idea of generating lift forces. Instead, the trend is toward a long straight-line pull–push action. Drag propulsion is again seen as the dominant propulsive force. Depending on the angle of the swimmer's hand (and other propulsive surfaces like the forearm), the powerful propulsive force of drag combines with lift to produce a resultant force that aids in propulsion (see figure 10.21).

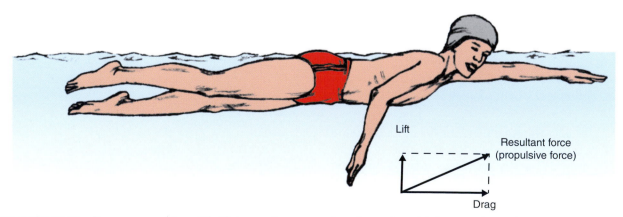

FIGURE 10.21 Drag can combine with lift to produce a resultant force that aids in propulsion during certain phases of hand and arm motion.

Videos of today's high-performance swimmers show that most use a predominantly straight-line pull–push pattern, particularly in freestyle, backstroke, and butterfly. In the breaststroke, drag forces are also considered the dominant propulsive force even though this stroke demonstrates the greatest amount of lateral hand and arm motion. The following list summarizes what swimmers should do as suggested by modern trends in freestyle (for an illustrated sequence of the freestyle stroke, see chapter 13):

- Work to improve the efficiency of propulsion and to reduce drag forces to a minimum. Keep your body stretched out and in a horizontal position to minimize form drag, frontal resistance, and wave drag.

- Gain a feel for the water by swimming with long, relaxed, rhythmic actions. Train to achieve long-range strokes with each arm.

- Avoid a high-cadence "water-pounding" arm action. This technique forces you to shorten your stroke, and you will run out of energy. Research indicates that energy consumption in water increases according to the cube of the stroke rate. In other words, doubling the cadence of your arm action through the water demands eight times ($2 \times 2 \times 2$) the energy.

- Avoid overemphasizing lateral (out to the side) motions and vertical (deep extended) motions with the hand and arm in the propulsive phase of the stroke.

- When the arm is pulling under the water, flex at the elbow. Remember that the forearm is important as a propulsive surface and that the hand is a mobile extension of the forearm.

- Pull and push against the water for the longest possible time along a line that is predominantly parallel to the direction of swim. For constant speed, make sure that as one arm pulls, the other is recovering in a fluid, cyclic manner.

- Keep your head down because doing so will raise your legs into an efficient drag-reducing position. Roll with each stroke so that you continuously present the narrowest possible profile to the water.

- When your hand enters the water ahead of your body, catch the water and try to use your momentum to move forward, past the hand position where you have caught the water.

- Eliminate jerky up–down, side-to-side motions that generate the massive resistance of wave drag. Water rewards smooth fluid actions and punishes swimmers who think that they can smash their way through the water from one end of the pool to the other.

- Consider shaving your skin where it comes in contact with the water and wearing a full-length bodysuit, especially if your leg kick is weak and not doing its job in maintaining a horizontal position in the water. A full bodysuit (with no wrinkles) increases buoyancy and helps hold your body in an efficient drag-reducing position.

Lift and Vortex Propulsion

A **vortex** is a mass of swirling fluids or gases. Vortexes (also called vortices) occur at the wingtips and trailing edges of airplane wings. They also occur spiraling away from paddles, oar blades, the hulls of boats, and the bodies of swimmers moving through water. In the butterfly stroke, a swimmer creates a vortex by simulating the undulating tail-lashing motion of a dolphin.

Modern research suggests that water that is spun into a vortex by the movement of a swimmer is given rotary inertia and has the tendency to resist lateral movement; instead, it rotates in the same spot, much like a child's spinning top. The undulating motions of a swimmer, particularly in the butterfly stroke, generate these vortices; and a swimmer can push against them to move through the water (see figure 10.22). Each spinning vortex can be viewed as a post against which the swimmer

FIGURE 10.22   Vortex propulsion in the butterfly.

can apply and exert force. Viewed from the side, the swimmer's actions look like those of a fish as seen from above. The athlete moves through the water, setting up vortices both above and below the body and then thrusts against these vortices. In the 2016 Rio de Janeiro Olympic and Paralympic Games, the undulating motions that generated vortices were used by most swimmers at the start of races in strokes other than the butterfly, particularly when the swimmers were temporarily below the surface of the water.

In addition, in swimming, vortex generators come into play. Vortex generators are raised ridges of various shapes and sizes that are positioned across the chest, lower back, and buttocks of full-body and partial-body swimsuits. They reduce both form and surface drag as the swimmer speeds through the water. Vortex generators on swimsuits act like the dimples on a golf ball. In the case of the swimmer, they provide the greatest benefit when the athlete is moving in a streamlined extended body position below the surface of the water. They are particularly effective when the swimmer is submerged and moving underwater—at the start of a race, after turns at the end of the lap, and during the underwater drive to the wall at the finish.

Vortex generators cause spiraling masses of water to flow parallel to the long axis of the swimmer's body. They mix faster-moving streams of water that occur outside the boundary layer with the slower-moving boundary layer water that passes in contact with the swimmer. The objective of the mixing action is to delay the breakup of the boundary layer of water flowing over the swimmer's body.

Magnus Effect

The **Magnus effect** (often called the **Magnus force** and named after its 1852 discoverer Gustav Magnus) is a lift force of tremendous importance to all athletes who want to bend the flight of a ball. You see the Magnus effect at work in the curved flight path of balls that are thrown, hit, or kicked and at the same time are given spin. Golfers, baseball pitchers, and soccer, tennis, and table tennis players all employ this effect to curve the flight path of the ball. The game of baseball in particular is made more fascinating by the Magnus effect. The ability of a pitcher to use spin when throwing curveballs, sliders, and screwballs and to minimize spin when pitching a knuckleball—and then have a batter hit those pitches—is the essence of baseball.

The Magnus effect operates in the following manner. Pitchers create topspin when they throw a baseball that rotates in a counterclockwise direction and at the same time travels forward, from right to left, as shown in figure 6.11 in chapter 6. The combination of counterclockwise rotation and forward travel means that the air particles on the top of the ball are traveling more slowly than the air particles on the bottom because on the top of the ball the counterclockwise spin is opposite to the direction of travel. The slower velocity of air on the top of the ball creates a high-pressure area, and the faster velocity on the bottom creates a low-pressure area. This pressure differential causes the path of the ball to move from the high-pressure area to the low-pressure area and (seemingly) magically creates the curved path of the ball, the phenomenon known as the Magnus effect (see figure 10.23).

The Magnus effect can be applied in any direction, and in this way an athlete can create backspin, topspin, and sidespin.

- Soccer players, for example, are well known for the way they use "banana kicks" (i.e., the Magnus effect) to curve free kicks and corner kicks around defenders and into the goalmouth.

- The athlete is simply creating torque on the ball, causing a pressure differential that alters the path of the ball.

- Tennis players and volleyball players use the Magnus effect when they apply topspin to make the ball drop suddenly while in flight.

- Elite golfers apply Magnus forces to produce draws and fades, and the weekend

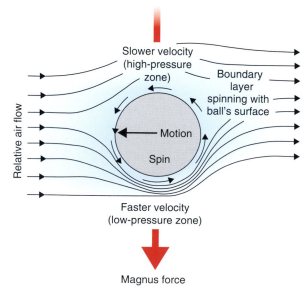

FIGURE 10.23   The Magnus effect on a ball.

hacker unwittingly applies spin and uses the Magnus effect to slice the ball off to the left and right.

- The Magnus effect can combine with the force of gravity or fight against gravity.
- A topspin combines with the downward pull of gravity, so topspin forehands in tennis (and table tennis) arc viciously over the net and down toward the court.
- A topspin rotates in the same direction that the ball is traveling, and the spin causes the ball to accelerate after it hits the court surface.
- A backspin, on the other hand, fights against gravity. The more spin that is present, the more the ball will hang in the air.
- Because the backspin is rotating in the direction opposite to that of ball travel, the spin causes the ball to slow down and even jump backward when it hits the court surface.
- Experienced players are able to read the spin on the ball from the motion of their opponent's racket.

A golf club like an eight or nine iron is steeply angled to give the ball tremendous backspin. The spin helps the ball fight gravity and gives it terrific lift, as well as the possibility of stopping dead or rolling backward after it lands. The raised stitches on a baseball produce the same effect as dimples on a golf ball. The seams grab a thick boundary layer, so a spinning ball gets plenty of help from the Magnus effect. Pitchers throw curveballs with a powerful snapping action of the wrist, which gives the ball terrific spin. The more spin that the ball has, the greater the Magnus effect is and the more the ball curves. Pitchers frequently combine topspin and sidespin so that the ball not only drops but also moves laterally across the plate, as shown in figure 10.24. Spin, gravity, and drag forces all work together to produce this effect.

What happens when hardly any spin is put on a baseball? In baseball, a pitch with slight spin is a knuckleball; in volleyball, a serve with hardly any spin is a floater. The characteristics of a knuckleball and a floater can be summed up in one word: unpredictable. In baseball, not even the pitcher knows for sure where the ball is going to end up. Both the knuckleball and the floater shift and flutter around in flight. This erratic movement is confusing to the batter in baseball and to the receiver in volleyball. A pitcher gives a knuckleball a lobbing or pushing release so that it travels slowly

FIGURE 10.24 A pitch, influenced by the Magnus effect, that moves down and laterally across the plate.

and spins maybe one full revolution by the time it reaches the batter.

- During flight, air flowing past at one instant grabs at the seams and at another instant contacts the smooth surfaces of the ball.
- The ball may go straight for a while and then suddenly veer to the right or left and possibly back again.
- Pitchers have learned that a ball released at a certain speed will first demonstrate a regular flight pattern.
- Halfway to the plate the ball slows down to a critical level, at which point drag forces build up dramatically and the ball suddenly drops.
- All this is meant to confuse the batter. But if the pitcher makes the error of throwing the ball with too much spin or too much speed, a knuckleball becomes an easy target for the batter.

Factors That Influence Moving Through Air and Water

Air is obviously not as thick, dense, or sticky as water, but both air and water vary in density and consistency. Pressure variations, temperature changes, and differences in what air and water contain (like water droplets in the air and salt in water) all change the way these fluids act and the way they affect sport performances.

Athletes compete when it's hot or cold, in locations from sea level to high altitudes, and at times when humidity (i.e., the percentage of water vapor in the air) is high. Variations in these conditions affect how fast athletes move, how far baseballs and badminton shuttles fly, and how much movement can be put on a curveball or knuckleball. The following summary of what happens with variations in air temperature, pressure, and humidity should be useful:

- *Air temperature.* When air temperature increases, its density decreases and its resistance against a moving object decreases.
- *Barometric pressure.* If barometric pressure (i.e., air pressure) decreases (as you go from sea level to higher altitudes), the density of air decreases and its resistance against a moving object decreases.
- *Humidity.* When humidity increases, the density of the air decreases and its resistance against a moving object decreases.

Of course, the drag-reducing effect of humidity doesn't apply when humidity switches to precipitation. In a downpour, water is no longer in a vapor state, and a baseball thrown by a pitcher has to push its way through the rain. In this situation the ball must contend with the considerable resistance of falling rain.

These atmospheric characteristics tell us that an athlete or an object (e.g., a baseball or a golf ball) will travel faster and farther in warm conditions than in cold, faster and farther at high altitudes than at sea level, and faster and farther on a humid day than on a dry day. For this reason, batters hit balls farther and pitchers hurl fastballs faster in Denver, where the ballpark is well above sea level.

On the other hand, pitchers hurling curveballs and knuckleballs need thick air that grabs at the seams of the ball and helps move the ball around. What's good for a fastball doesn't produce the fluttering deception of a knuckleball. All the pitches thrown in baseball are affected in one way or another by environmental conditions. As a generalization, dense air helps move the ball around, whereas thin air at high altitudes makes it easy for the ball to go faster. Knuckleballs become less deceptive at high altitudes and as temperature rises. In these conditions, the game belongs to the slugger and the fastball pitcher.

Just as a fastball travels faster at high altitudes, so will a sprint cyclist. Some of the fastest times recorded by competitive cyclists, from the 200 m sprint to the 1 h time trial, have been set at high-altitude venues such as Colorado Springs; Bogota, Colombia; and Mexico City. In the 1 h time trial, the gain from cycling at high altitude can be partially offset by the reduced oxygen available for the athlete. To counteract this problem, prior conditioning at high altitude is a necessity. The benefits provided at high altitudes for a fastball pitcher, a home run hitter, and a sprint cyclist can cause difficulties for badminton players. A badminton shuttle hit with the same force and trajectory at high altitudes as at sea level will travel farther because resistance to its flight is lower. To counteract this problem, these badminton shuttles weigh less than those used at sea level, allowing the thinner air at high altitude to reduce their velocity.

The consistency of water differs just as it does for air. Water varies in density depending on its chemical makeup. An athlete attempting to swim in the Dead Sea, situated on the Jordan–Israel border, will be shocked at the difference between the water there and the water in a swimming pool. The Dead Sea is incredibly thick and dense because of its high salt content. This characteristic dramatically increases its drag. But the water offers a benefit! Even the leanest and most muscular athlete can lie back in the saline water of the Dead Sea and, with little or no movement, be sufficiently buoyant to read a book without fear of sinking.

AT A GLANCE

Factors That Influence Flight

- To generate lift in flight, the athlete (or object) can be angled or tilted relative to the direction of fluid flow passing over it.
- The Magnus effect is a lift force of tremendous importance to all athletes. Creating spin on an object as it passes through a fluid can generate a difference in velocity around the object, which in turn causes a pressure differential and ultimately the Magnus force that changes the path of the object.
- The Magnus effect can combine with the force of gravity or fight against the force of gravity.

SUMMARY

- Going with the flow requires the concept that a fluid (air or water) will flow around an athlete and that any fluid particles that were separated will eventually reconnect.
- An athlete or object moving through a fluid is affected by the mechanical principles of hydrostatic pressure (exerted by the weight of a fluid), buoyancy (the force opposing gravity that acts on objects partially or totally immersed in a fluid), drag (the force opposing motion through a fluid), and lift (the force acting perpendicularly to motion that deflects an object from its original pathway). The influence of flow is varies depending on whether the movement is in air or water.
- The pressure that the atmosphere exerts on the earth's surface is 100 kPa (14.7 psi). An increase in altitude decreases this pressure.
- Water is virtually incompressible, and consequently the pressure exerted by water increases with depth. In the ocean, pressure increases approximately one atmosphere (100 kPa) for every 10 m (33 ft) of depth.
- The buoyant force acting on an athlete or an object is equal to the weight of the fluid that the athlete or object displaces when immersed in the fluid.
- The center of buoyancy is the place where the buoyant force concentrates its upward thrust on an object immersed in a fluid. The center of buoyancy is usually positioned higher on an athlete's body than the athlete's center of gravity. The center of buoyancy of an object or athlete is in the same place as the center of gravity of the displaced water.
- Atmospheric pressure, temperature, and the contents of air or water affect how water and air act. The more dense or viscous a fluid is, the greater the frictional forces acting on an athlete or an object passing through that fluid are.
- Buoyancy plays an important role in sport, particularly in relation to the differences in hydrostatic pressure generated from salt water or freshwater.
- When moving through a fluid, the individual forces of drag and lift combine to produce a resultant force. The drag forces are in the direction of travel and oppose, or drag against, motion.
- The three types of drag are surface drag, form drag, and wave drag.
- Surface drag is also called skin friction or viscous drag. The amount of surface drag is determined by the relative motion of object and fluid, the area of surface exposed to the flow, the roughness of the object's surface, and the fluid's viscosity. Surface drag increases according to the square of the velocity.
- Form drag is also called shape drag or pressure drag. The amount of form drag is determined by the relative motion of object and fluid, the pressure differential between the leading and trailing edges of the object, and the amount of surface acting at right angles to the flow. Form drag increases according to the square of the velocity.
- Wave drag occurs at the interface between water and air. The amount of wave drag is determined by the relative velocity at which the object and wave meet, the surface area of the object acting at right angles to the wave, and the viscosity of the fluid. Wave drag increases according to the cube of the velocity.
- At high velocities, turbulent flow produces a low-pressure wake acting to the rear of an object or an athlete. This low-pressure area is used in sport for drafting, or slipstreaming.
- Balls moving through the air are strongly affected by form and surface drag. Increasing the surface drag can reduce the lower-pressure wake to the rear of the ball, and an increase in surface drag can help decrease the ball's form drag. Dimples on a golf ball increase its surface drag.
- Athletes and objects are affected by lift forces that depend on the relative motion of the object and the fluid, the angle of the object relative to the flow of the fluid, the size of the surface area angled into the fluid flow, and the nature (e.g., density) of the fluid.

- Swimmers angle their hands and feet to create lift, which can act as a propulsive force. Modern research shows that the greatest propulsive force for a swimmer comes from pulling and pushing back against the water as long as possible in a direction parallel to the long axis of the swimmer's body. This propulsive force is called drag propulsion.

- Vortexes (or vortices) are masses of swirling water produced by undulating lashing motions of a fish's body and tail. Athletes mimic this action, particularly in the dolphin kick of the butterfly stroke. Vortices generated in this manner are considered to reduce drag and assist in propulsion.

- Vortex generators are a series of raised ridges placed across the surface of full-body and partial-body swimsuits. They are intended to reduce the form and surface drag of the swimmer, particularly when the athlete is moving below the surface of the water.

- The three types of drag forces (form, surface, and wave) can be addressed in sporting applications through modification of the equipment interface, athlete technique, or both.

- A spinning object (e.g., a ball) traveling through the air builds up high pressure on the side spinning into the airflow. Low pressure occurs on the side spinning in the same direction as the airflow. The ball is deflected from high pressure to low pressure. This phenomenon is called the Magnus effect.

KEY TERMS

airfoil	drag	relative motion
angle of attack	form drag	surface drag
Archimedes' principle	hydrostatic pressure	turbulent flow
Bernoulli's principle	laminar flow	viscosity
boundary layer	lift	volume
buoyancy	Magnus effect	vortex
buoyant force	Magnus force	wake
center of buoyancy	propulsive drag	wave drag

REFERENCES

Asai, T., K. Seo, O. Kobayashi, and R. Sakashita. 2007. "Fundamental Aerodynamics of the Soccer Ball." *Sports Engineering* 10 (2): 101-109.

Formosa, D., M.G.L. Sayers, and B. Burkett. 2014. "The Influence of the Breathing Action on Net Drag Force Production in Front Crawl Swimming." *International Journal of Sports Medicine* 35 (13): 1124-29.

Hemelryck, W., P. Germonpré, V. Papadopoulou, M. Rozloznik, and C. Balestra. 2014. "Long Term Effects of Recreational SCUBA Diving on Higher Cognitive Function." *Scandinavian Journal of Medicine and Science in Sports* 24 (6): 928-34.

Jinji, T., and S. Sakurai. 2006. "Baseball." *Sports Biomechanics* 5 (2): 197-214.

Pelz, P.F., and A. Vergé. 2014. "Validated Biomechanical Model for Efficiency and Speed of Rowing." *Journal of Biomechanics* 47 (13): 3415-22.

Swann, C., L. Crust, and J. Allen-Collinson. 2016. "Surviving the 2015 Mount Everest Disaster: A Phenomenological Exploration Into Lived Experience and the Role of Mental Toughness." *Psychology of Sport and Exercise* 27:157-67.

 Visit the web resource for review questions and practical activities for the chapter.

PART II

Putting Your Knowledge of Sport Mechanics to Work

11

Analyzing Sport Skills

Oleg66/E+/Getty Images

When you finish reading this chapter, you should be able to explain

- how to determine skill objectives;
- how knowing the special characteristics of a skill can help you analyze athletic performance;
- how to create a checklist for analyzing a sport;
- what you gain from an analysis of the performances of elite athletes;
- how technology can help you analyze sport skills;
- how to divide a skill into phases and key elements; and
- how to use your knowledge of mechanics in the analysis of a skill and the correction of errors.

The next three chapters (chapters 11, 12, and 13) pull together the fundamental knowledge that has been covered in this text and puts these learnings into practice, which is the essence of applied sport mechanics. Chapters 11 and 12 are closely linked and will show how to put your knowledge of sport mechanics to work. Here you'll find advice on how to break a skill into smaller parts. This process will make it easier when you critically observe an athlete's performance. Chapter 11 gives examples of observation techniques and teaches you how to select errors that need correcting.

One of the greatest challenges you'll face working in sport is watching the athlete perform and deciding which aspect of the skill needs correction (if any). After you have analyzed the performance, you need to communicate this information to the athlete. Technology such as the video recording we described in the earlier chapters can be a real asset here. The video image allows the athlete to see what you are talking about. In addition, if you take measurements of the performance, such as a running cadence, you can compare this variable after you have made the correction or intervention. These objective data will allow you as a sport scientist to provide evidence that can persuade the athlete to make a correction in technique—which, after all, is what sport mechanics is all about.

If you don't have a well-planned approach, you're likely to be overwhelmed by the complexity and speed of the skill you are trying to analyze. You won't know what aspect of the skill to look at or what error to correct first. In fact, you may see so many errors at once that you throw your hands up in the air and in desperation give vague sport science feedback such as "Hit harder" or "Be more aggressive!" Advice like this is of little help to the athlete because people understand cues in different ways. Any advice provided must be clear and to the point so that the athlete will understand exactly what is required.

What you need to do is gather background information about the skill beforehand and come to each training or exercise session with a precise plan to guide your observation, your analysis, and your correction of errors. If you understand the mechanics of the skill that the athlete is performing and know how to go after major errors, the athlete or coach benefits immensely and quickly improves in performance. The following steps provide the information you need before you start correcting errors:

Step 1: Determine the objectives of the skill.

Step 2: Note any special characteristics of the skill.

Step 3: Study top-flight performances of the skill.

Step 4: Divide the skill into phases.

Step 5: Divide each phase into key elements.

Step 6: Understand the mechanical reasons that each key element is performed as it is.

If you work your way through each step, you'll learn how to break a skill into important parts (or phases), and you'll know how to use your knowledge of sport mechanics when you analyze each phase (Travassos et al. 2013). You'll find out how much easier it is to analyze each phase of a skill separately rather than evaluate the total skill and then try to recollect what happened.

Don't think you must go through each step every time you teach or assess a sport skill. After you have read this chapter, you'll understand what information you need, and with a little practice you'll be able to carry out most of the steps in your head. To begin with, however, write down on a clipboard the information required. Then take this material with you and use it as a guide during your training or testing sessions.

Step 1: Determine the Objectives of the Skill

The rules of the sport and the conditions that exist when a sport skill is performed determine **skill objectives**. Most skills have more than one objective. Being aware of these objectives is helpful because they determine the technique and mechanics that your athlete must use to perform the skill successfully. Look at the flowchart in figure 11.1. At the top is the desired result, or skill; then we look at the objectives that can influence how this result, or skill, is achieved.

The most common objectives, in no particular order, of a sport skill are

- speed (for the athlete or object to travel as fast as possible),
- accuracy (for the athlete or object to move as accurately as possible),
- form (for the athlete to move with the precise positioning needed), and
- distance (for the athlete or object to travel as far as possible).

Let's look at a few sport skills to illustrate what we mean by skill objectives. To run the 100 m, the key objective is speed. Olympic competitors are not worried about how accurate they are (as long as they stay in their lane); they just want to travel as fast as possible. If the athlete is a marathon runner, form may be more important than it is for the 100 m runner because in a marathon, form can influence energy expenditure.

The dominant objective for an athlete competing in the discus event is to throw the implement as far as possible. The farther the discus travels, the better the performance is. But the discus must

land within a sector, so accuracy of flight is an important objective as well. The distance thrown is not counted if the discus lands outside the sector lines. The underlying objectives provide the most important aspects of this sport:

- first, accuracy (so that the discus lands within the sector) and
- second, distance (so that the discus travels as far as possible).

In addition, if the thrower loses balance and falls out of the ring, the throw is declared invalid even if the discus lands within the sector lines.

The objectives of accuracy and distance determine what applied sport mechanics principles to keep in mind when working with the coach and athlete in the sport of discus (Van Biesen et al. 2017).

- The overriding importance of accuracy tells you that the technique is important.
- To achieve distance, the mechanical objective in the event is maximum velocity at release.

Therefore, you should concentrate on teaching the athlete or coach how to make the discus leave the throwing hand as fast as possible. How the discus leaves an athlete's hand and how it spins determine its flight characteristics and its distance (again, these sport mechanics principles are important). So you cannot forget that an optimal spin and trajectory are important objectives, too. Remember as well that the body positions that the athlete uses during the throw influence the distance and flight of the discus and the athlete's stability after the discus is released. It would be heartbreaking if an athlete threw a world-record distance only to have a foul declared because she fell out of the front of the ring or stepped on the rim of the ring during the throw.

In a volleyball spike, an athlete must jump high enough to strike the ball over, around, or off the blockers. The prime objective of a spike is to make the ball hit the floor in the opponent's court (Van

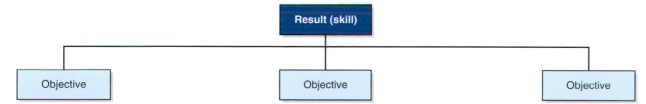

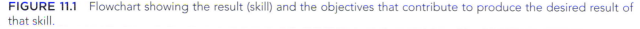

FIGURE 11.1   Flowchart showing the result (skill) and the objectives that contribute to produce the desired result of that skill.

Biesen et al. 2017). To achieve this objective the key aspects are

- jumping ability and timing,
- accuracy in directing the ball, and
- taking care not to contact the net.

Keep these objectives in mind when working with the coach and athlete during the spiking skills. Work on the mechanics of the approach, the jump, and the spiking action and then on control of the body after the ball has left the athlete's hand.

Compare the objectives of the volleyball spike with those required of a high jumper. Height is obviously a prime objective in high jump just as it is in a volleyball spike. But a high jumper is also required to cross a bar, an objective not required of a volleyball player. So a high jumper needs to

- jump both vertically and horizontally,
- rotate in the air to get into a good bar clearance position, and
- maintain good form to avoid knocking the bar off on the way up or down.

In Olympic weightlifting, the prime objective of both the clean and jerk and the snatch is to hoist a barbell to arm's length above the head. A secondary objective is to demonstrate control over the barbell after it's in this position. This second criterion is necessary to ensure that the judges pass the lift. Even though the barbell must be held steady for a relatively short time, control and stability are important objectives that must be taught for the athlete to achieve success in this skill.

Whatever sport you are working with, whether an individual or team sport, be aware of all the objectives required of each skill. If you focus on satisfying one objective and forget or deemphasize another, you'll limit the success of the athlete.

- What use is it if a water polo player learns to fire the ball at phenomenal velocity if he cannot control and direct the path of the ball?
- Similarly, what use is it if you teach a diver how to get great height and spin if the entry into the water is a disaster?

You need to be aware of all the objectives required by a skill and remember that all these objectives play a part in determining the technique that you teach the athlete.

An important step you can take as an applied sport scientist is to develop a checklist of skills for a particular sport (Young et al. 2014). Many references are available to establish this checklist, such as the snapshot in table 11.1. For this checklist, the assessor inserts skills specific to the sport in the first column and then tests each athlete on these skills. Coaches for specific sports can develop their own checklists by combining the mechanical components in the sport (discussed in earlier sections of the text) and then applying this knowledge to the specific sport they are coaching. These checklists are often developed and used to provide a more scientific approach to analyzing sport skills.

Step 2: Note Any Special Characteristics of the Skill

Sport skills can be divided into different types based on the manner in which athletes perform the skill and the conditions under which they perform the skill. Manner and conditions are interrelated, and both dramatically influence the methods you use when assessing the performance. For example, if you consider the manner in which skills are performed, you'll see that some skills are performed once and then a totally different action occurs.

TABLE 11.1 Snapshot of a Skills Checklist

Skills identified	Objectives	Skill rating				
		Weak				Strong
Technical skills						
Skill 1:		1	2	3	4	5
Skill 2:		1	2	3	4	5
Skill 3:		1	2	3	4	5
Skill 4:		1	2	3	4	5
Skill 5:		1	2	3	4	5
Skill 6:		1	2	3	4	5

Reprinted by permission from R. Martens, *Successful Coaching*, 4th ed. (Champaign, IL: Human Kinetics, 2012), 206-207.

Other skills are different because they repeat cyclically (i.e., over and over). These two types can be called **nonrepetitive skills** and **repetitive skills**. The conditions under which athletes perform skills also differ considerably.

- Some conditions are controlled and predictable; you know what the conditions will be before the competition starts.
- Other conditions vary considerably and are unpredictable, and knowing what they'll be when the competition begins is difficult.

The following sections cover nonrepetitive skills, repetitive skills, predictable environments, and unpredictable environments.

Nonrepetitive Skills

Nonrepetitive skills are often called **discrete skills** because they have a definite beginning and end, even though they can be performed more than once in a sporting situation. Here are a few examples:

- Tower dive
- Shot put
- Baseball bunt

Skills such as these are not repeated in a cyclic pattern. Instead, some other action occurs immediately afterward. A diver lands in the pool, climbs out, and waits for a turn in the next round of dives.

A similar situation occurs for the shot putter, who after throwing must wait for other competitors to complete their throws before performing again. The baseball player follows a bunt with a completely different action, usually a sprint to first base. You can easily teach nonrepetitive skills as separate entities. After the athlete has learned and mastered the skill, you can add some other skill or action to lead into it or lead out of it—similar to what occurs when the baseball player bunts and sprints to first base or the gymnast performs a handspring followed by a dive roll.

Frequently, the impetus generated in one nonrepetitive skill carries over and assists in beginning another nonrepetitive skill:

- A young gymnast builds a floor exercise routine in this way. A front handspring may join a front somersault, and the somersault may lead into another skill.
- Similarly, a triple jumper hops, steps, and finally jumps. The three jumps differ, yet the skill of triple jumping depends on the synchronization of all three skills. For an excellent distance, the hop must contribute to the step and the step must add to the jump (see figure 11.2).

When you work with athletes performing nonrepetitive, or discrete, skills in sequence, you need to assess each skill separately. Then you can teach the athlete to adapt to the rhythm pattern and changes that occur when two or three skills are performed in sequence. Be aware that performing two or three skills in sequence presents additional difficulties to the athlete. Novice triple jumpers frequently perform an immense hop only to collapse at the end of it and have nothing left for the step or the jump. They do not produce a balanced effort or flow from the hop to the step and finally to the jump. In gymnastics, a young athlete can learn to perform a back somersault by itself. Then you can teach the performance of a round-off that leads into the back somersault. If correctly performed, the round-off makes the performance of a back somersault easier. Performed poorly, the round-off positions the gymnast incorrectly for the takeoff into the back somersault, making it difficult for the gymnast to get around and safely complete the somersault.

FIGURE 11.2   The triple jump is an example of three discrete skills in sequence: the hop, the step, and the jump.

APPLICATION TO SPORT

How Do Elite Athletes Compare Athletically With Other Species?

Although our best athletes produce great performances, those efforts don't compare with what animals and insects can do! When searching for the best rhythm in a sport skill, the answers can sometimes be found by observing the skills from the animal kingdom. In a sprint, a cheetah can finish the 100 m while an Olympic sprint champion is still accelerating at the 30 to 40 m mark. Kangaroos jump 2.7 m (9 ft) vertically with ease, and impalas have no trouble leaping 12 m (40 ft) horizontally. Fleas jump more than 150 times their own length, vertically or horizontally; to match this performance, athletes would have to

© Royalty-Free/CORBIS

jump close to 300 m (1,000 ft)! These achievements are a combination of the unique anatomy of the animal, as well as the technique they use to execute the skill. Relative to body weight, the best Olympic weightlifters in the world are easily outmatched by ants, which can carry in excess of 50 times their own body weight. In water, the top speed of around 13 km/h (8 mph) by our best swimmers is five times slower than that of the bluefin tuna, which has hit speeds of 83 km/h (45 knots)!

Repetitive Skills

Repetitive skills have a cyclic, continuous nature. For example, the actions that make up the movement pattern of sprinting are repeated continuously during a race. This repetitive, continuous feature occurs in many sports, such as

- race walking,
- cycling,
- swimming,
- speed skating, and
- cross-country skiing.

The most important aspect of repetitive skills is that one complete cycle of the skill immediately leads into the next. A **follow-through** (which slows down and dissipates energy in a nonrepetitive skill) becomes a recovery in a repetitive skill and is essential for maintaining continuity and rhythm.

In competitive swimming, athletes aim for a fast arm recovery when they perform their strokes. The arms complete their pull in the water and then quickly cycle forward into the next propulsive action. No braking action or dissipation of energy occurs as it does in the follow-through of a discus or javelin throw. Like a cyclist who wishes to keep the pedals spinning at a high rate, a competitive

swimmer wishes to do the same thing with the arms after each arm pull (see figure 11.3).

Repetitive skills are frequently taught to young athletes in much the same way as nonrepetitive skills. The freestyle stroke is broken down into leg action, arm action, and breathing. These components of the stroke are taught separately and then molded together to build the complete skill. The number of repeats, or cycles, of the total skill is progressively increased as the athlete's ability improves.

Skills Performed in Predictable Environments

Some sport skills are performed in a precise and predictable environment. These types of skills are frequently described as **closed skills**. In this situation, the athlete can get on with the job of performing the skill without having to make quick decisions because of a sudden change in conditions. A clean and jerk in weightlifting and the skills in a synchronized swimming routine are examples. The fact that the athlete can concentrate on the lift or on the skills in the routine, without worrying about the actions of opposing players or changes in weather, makes practice sessions easier to plan and training easier for the coach and athlete.

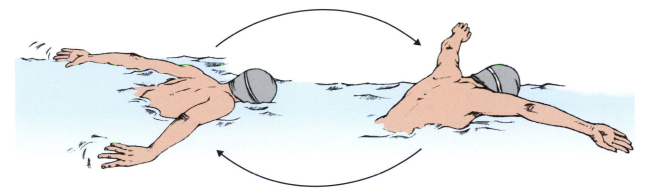

FIGURE 11.3 Swimming strokes are examples of cyclic, repetitive skills in which the recovery of the arms and legs leads into the next propulsive phase.

Skills Performed in Unpredictable Environments

On the other hand, some sport skills are performed in an unpredictable environment. These skills are often described as **open skills**. The most frequent cause of an unpredictable environment is the presence of opposition whose prime purpose is to make athletes fail in whatever they are trying to do. Consequently, athletes must respond according to the conditions that occur in any instant during the competition:

- In baseball the batter responds (in less than half a second!) to whatever pitch is thrown.
- In volleyball the player responds according to the serve that comes over the net. The response is going to be different for a floater serve than for a fast topspin spike serve.
- In freestyle wrestling and judo the athlete attacks or defends according to the maneuvers of the opponent.
- In soccer a goalkeeper reacts according to the maneuvers and the shot made by an attacking player.

Wind, waves, rain, sun, and varying field and court conditions can also cause uncertainty and unpredictability. A surfer must assess the nature of the wave and perform surfing skills accordingly. Each wave needs to be considered individually when it occurs, and the surfer must develop an ability to cope with these conditions. The variability that exists in all the sports we've mentioned, from baseball to wrestling, forces the athlete to make sudden decisions and to perform skills at varying velocities. The ability to judge the situation and to react quickly is obviously an important element of success.

When you work with sports that are open skills, which are performed in unpredictable conditions, begin by making the situation as predictable as possible.

- Wrestlers work repeatedly on the same defensive maneuver against an opponent who is required to repeat the attacking move.
- In baseball and tennis, players face balls fired repeatedly and predictably from pitching and serving machines.
- In rugby, football, and field hockey, athletes practice set plays without opposition. Then other team members work as opposition, and the same plays are repeated.

In this way the mechanics of a particular skill are practiced in a predictable situation until the quality of the skill performance is acceptable. Then more unpredictability is introduced.

How soon unpredictability is introduced depends on many factors, one of the most important being how fast the athlete learns the required skill (Farrow and Robertson 2017). Many coaches like to move quickly to unpredictable situations. Others mix it up so that in some drills the athlete learns rapidly how to judge what should be done, and in other drills the athlete works on a particular skill repeatedly under predictable conditions.

Step 3: Study Top-Flight Performances of the Skill

Throughout their sporting careers, all athletes can watch high-performance athletes perform. This observation doesn't necessarily have to be step 3 in the sequence as described in this chapter. But watching the best perform a skill or event is certainly worthwhile.

- When you watch top-class athletes perform a skill, you get a picture of the speed, rhythm, power, body positions, and other characteristics that make up a quality performance.

- This observation helps you understand the basic movement patterns in the skill that you need to focus on.

- Use a video camera to tape these performances from various angles. Then you can watch the skill repeatedly at normal speed and in slow motion.

You'll soon notice that, in spite of differences in body type, the techniques that top athletes use all show common features:

- High-performance golfers shift their body weight and rotate their hips in much the same way.

- Great throwers in track and field use similar throwing positions and activate their muscles in a similar sequence.

- Top-class divers use a similar hurdle step, and they drive up off the springboard with similar arm and leg actions.

These identical features exist because top-class athletes use good mechanics. Their support staff (coaches and sport scientists) taught them to use actions in their performances that produce the optimal sport mechanics, namely force, velocity, spin, and so forth required by the skill.

As you progress through steps 4, 5, and 6 as explained in this chapter, you'll get used to associating sport mechanical principles with technique.

You'll start using your knowledge of mechanics when you look at a high-performance athlete, so that you can say to yourself, "I understand the mechanical reasons that these champion athletes shift their weight and rotate their hips when they drive a golf ball, and I understand why they extend their arms when the club head contacts the ball." You'll realize that these technical features are necessary actions that must be taught to all young golfers irrespective of their shape, size, and build.

The same principles apply to the skills of any sport. Elite performers use good technique based on sound mechanics and therefore provide you with a model on which to form your analysis of the sporting skill.

Step 4: Divide the Skill Into Phases

Your next task is to divide the skill you're working with into phases. Going through this process makes your job much easier when you look for errors in the athlete's performance. Quite simply, this method stops you from becoming confused by trying to watch too much of the skill at the same time.

Most skills consist of several phases. A **phase** is a connected group of movements that appear to stand on their own and that your athlete joins together in the performance of the total skill. Many skills, for example, can be broken down into the following four phases:

1. Preparatory movements (setup) and mental set

2. Windup (also called backswing)

3. Force-producing movements

4. Follow-through (or recovery)

If you look at a golf swing, a hockey slap shot, or a baseball pitch, preparatory movements and mental set make up the first phase in the skill, however brief they might be. The second phase consists of the **windup** (or backswing). The third phase comprises the **force-producing movements**. The fourth phase, the follow-through, completes the skill.

Each phase, starting from the preparatory movements and mental set, leads into and influences the next phase in line like a chain reaction, as shown in the golf drive example in figure 11.4.

- This common characteristic tells you that errors occurring during an early phase of

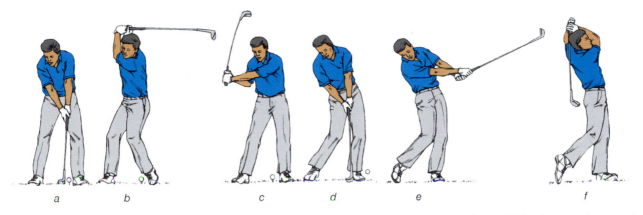

a   b   c   d   e   f

FIGURE 11.4   Phases of a golf drive are the (*a*) preparatory movements and mental set, (*b*) backswing, (*c* and *d*) force-producing movements, and (*e* and *f*) follow-through.

a skill are bound to affect all the phases that follow.

- Therefore, when something goes wrong at the end of a skill, examine not only the last phase but also earlier phases to see whether the root of the problem lies there.

For example, if a golfer makes an error in setting up and addressing the ball or performs the backswing incorrectly, the effect of the error carries into the remaining parts of the drive and, of course, into the flight of the ball. Don't be deceived into thinking that all errors stem from the phase in which they occur. Check out earlier phases—the problem often lies there!

Let's look at each phase individually to see what specific contributions each makes toward the performance of the total skill.

APPLICATION TO SPORT

Phases in Diving: How Divers Make Splashless Entries

At one time, a feetfirst entry was used in a dive from the 10 m (33 ft) tower. Divers found that if they flattened their feet on entry rather than pointed their toes, little or no splash occurred and the water simply bubbled at the surface. Similar experiments with headfirst entries led to divers clasping their hands and flattening and facing them toward the water. As with the flat-feet entry, this technique produced a low-pressure area that sucked the water downward behind the diver's hands and produced little splash. A problem with this technique was that the impact with the water could injure a young diver's wrists. To counteract this problem, young divers regularly wear wrist supports in training and occasionally in competition.

Steve Russell/Toronto Star via Getty Images

Preparatory Movements (Setup) and Mental Set

Preparatory movements and mental set include the motions and mental processes that the athlete goes through when setting up and getting ready to perform.

- A golfer takes up a stance and addresses the ball.
- A tennis player gets herself ready to serve and mentally decides where to direct the ball.
- An offensive lineman crouches with his muscles in a static stretch position. When the ball is snapped, his muscles respond with an explosive thrusting motion that immediately leads into the next phase of the skill.

Cyclic, repetitive skills may require preparatory movements at the start of the skill, after which these normally don't occur. For example, a butterfly swimmer doesn't establish a static stance before each propulsive action. The swimmer flows immediately from each arm pull and leg beat into the next.

Windup or Backswing

Many skills use a windup, or backswing, in preparation for the movements to follow. Whatever name is given to this phase, the objective remains the same—to stretch the athlete's muscles and establish a position from which she can apply force over an optimal distance or time (Ross et al. 2015). Examples include the following:

- The rotary windup of a discus thrower
- The backswing in golf and baseball
- The backward extension of a javelin thrower's arm

In a tennis serve and a volleyball spike, the dropping back of the hitting arm to the rear of an athlete's body fulfills a purpose similar to that used by a thrower. In kayaking, the forward reach of the paddler before the thrust of the blade into the water acts as a windup.

Force-Producing Movements

Force-producing movements are the specific actions that athletes use to generate force. They usually involve the athlete's whole body and may include an approach, but in finer, more discrete actions (such as archery or throwing a dart), they may require use of only the arm and shoulder muscles and minimally involve the muscles of the rest of the body.

Force-producing actions are tremendously important for creating the desired effect of a skill. The athlete's muscles need to apply force in the correct amount, over the correct range and time, and in the correct sequence. You'll find that force-producing actions come in many types. They include such sequential actions as

- the approach, pull, and push of the pole-vaulter;
- the body extension and arm flexion of the rower;
- the rotating spins and throwing actions of the hammer thrower and discus thrower; and
- the approach, takeoff, and arm actions of a basketball player performing a layup.

In contrast, in a powerlifter's deadlift, the force-producing actions occur almost simultaneously; the athlete's leg, back, arm, and shoulder muscles pull at the same time. In all skills, an important and critical instant in time occurs at the end of the force-producing movements. It happens when a baseball is struck, a takeoff occurs, or an implement is released. At this instant, the athlete has applied the optimal amount of force and set its direction. At this point, the athlete can do nothing more to upgrade the skill.

Follow-Through or Recovery

Follow-through or recovery actions occur immediately after the force-producing motions are complete. In throwing skills, the implement has been released, and in hitting skills, the impact has been made. In many skills coming to a complete stop immediately after completing the force-producing actions is impossible and even dangerous. The momentum generated causes the athlete's limbs to continue along their original pathway. The follow-through acts to dissipate the force of these actions safely.

In a swimming stroke, a skill in which the movement pattern is repeated in a continuous and cyclic fashion, the recovery of the arms leads quickly to the next repetition of the arm pull. In these repetitious skills, momentum and rhythm are an essential part of the cadence of the complete skill. The recovery actions help maintain balance and continuity of motion. Similar examples include the leg and arm recovery in sprinting, speed skating, and cross-country skiing.

Step 5: Divide Each Phase Into Key Elements

When you have chosen the most important phases of a skill, direct your attention toward the task of dividing each phase into its key elements. Key elements are distinct actions that join to make up a phase. Try to view a skill as a building you are erecting. Phases are the walls of your building, and the key elements are the bricks you use to make each wall.

How do you choose key elements? Identify the distinct actions that are essential to the success of each phase in the skill (the same way you identify elements that are essential to the success of the skill as a whole). A windup phase has its key elements, as does the force-producing phase and the follow-through. The following examples give you an idea of what key elements are, although we haven't listed every key element in the phases we've chosen.

You'll see these key elements in the techniques that all top-flight athletes use because they are essential for good technique and contribute mechanically toward the success of the skill. Without them, athletes could not produce optimal performances.

In the force-producing phase of a golf drive, the athlete shifts her body weight to the rear foot and from the rear to the forward foot. She rotates her hips into the drive and has extended arms when the club contacts the ball (see figure 11.4). The key elements in the golf drive are

- weight shift,
- hip rotation,
- head position, and
- arm extension.

In a high-jump approach, the athlete leans into the curved path of the approach, which is part of the force-producing phase of a high jump. At the completion of the approach he leans back and lowers his center of gravity when stepping into the takeoff position. The arms are positioned to the rear of the body in preparation for swinging forward and upward at takeoff (discussed in more detail in chapter 13). The key elements in the high jump are

- backward lean,
- lowering of the center of gravity, and
- placement of the arms to the rear of the body.

In the force-producing phase of a javelin throw, the athlete makes the approach, leans back, and steps forward into a wide throwing position. Next she rotates the hips and chest toward the direction of throw. Simultaneously, the athlete shifts her body weight from the rear to the forward leg (discussed in more detail in chapter 13). The key elements of the javelin throw are

- approach,
- backward lean,
- wide throwing stance,
- hip and chest rotation, and
- weight shift.

In a football punt, after stepping forward with the supporting foot, the athlete swings the kicking leg through a long arc. The kicking leg, which starts partially flexed, is fully extended on contact with the ball. The athlete simultaneously shifts his body weight forward and upward into the punt. His arms, which fed the ball onto the kicking foot, are extended sideways to maintain balance (discussed in more detail in chapter 13). The key elements of the football punt are

- extended base,
- weight shift,
- long kicking arc,
- leg extension, and
- arm extension.

Remember that each of these skills includes additional key elements and that the sequence in which these elements are performed is itself important. In some phases of a skill, key elements are performed almost simultaneously. In other situations, one element definitely flows to the next. With practice and careful observation of elite performances, you will be able to pick out all the key elements for each phase of a skill and understand the timing of their performance.

If we continue the theme of jumping from earlier in the chapter, you can see in figure 11.5 that several components contribute to the end result—a greater distance jumped. With all skills, some variables can be modified, such as position at takeoff in this example, and the coach can intervene and modify such factors. But other factors in sport skills are more permanent, such as the physique of the athlete in this example. By breaking down the skills of the sport, you can determine which components need finer tuning and which components are fixed. Your next job is to understand the mechanical reasons behind the key elements and what purpose they serve. This is the final step.

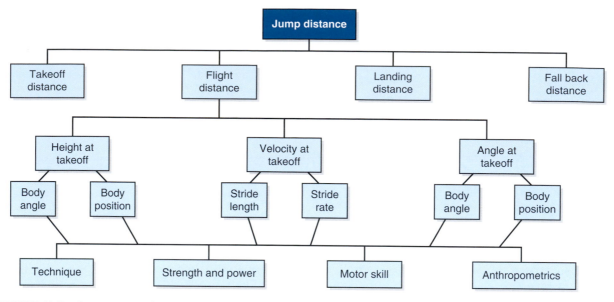

FIGURE 11.5   Components that contribute to distance in jumping.

Adapted from J.G. Hay, *The Biomechanics of Sports Techniques*, 4th ed. (Upper Saddle River, NJ: Pearson Education, 1993), 428.

AT A GLANCE

Phases of Sport Skills

- Most sport skills consist of several phases—the preparatory movements, windup (also called backswing), force-producing movements, and follow-through.

- Many skills use a windup, or backswing, in preparation for the force-producing movements to follow. This phase is important for creating the desired effect of a skill.

- Key elements are distinct actions that join to make up a phase essential to the success of the skill.

Step 6: Understand the Mechanical Reasons That Each Key Element Is Performed as It Is

Understanding the mechanical basis behind each key element is an important step in your sequence. The first 10 chapters laid the foundations of principles of sport mechanics. By analyzing technique we are putting this knowledge into practice—in essence, practicing applied sport mechanics.

All fundamental actions that an athlete takes within technique are founded on mechanical principles. In other words, technique is based on

mechanical laws. So after you've picked out the key elements in the skill you are analyzing, you have to understand the mechanical purposes behind each element. You must be able to answer questions such as the following with responses like the ones listed here.

Why cock and uncock the wrists during a golf drive?   Cocking and uncocking the wrists during a golf drive causes the golfer's arms and club to simulate the whiplash, or flail-like, action of the high-speed tip segments of a whip (see figure 11.4c-11.4e). When the wrists are cocked and uncocked, they act as an additional axis around which the club can rotate. The velocity developed from the swing (and length) of the golfer's arms is multiplied along the length of the club shaft (Fedorcik et al. 2012). Without the cocking and uncocking action, the arms and club would move as a fixed unit. This action would not allow the head of the club to reach optimal velocity.

Why should a sprinter's legs and arms thrust and swing parallel to the direction of sprint during a 100 m sprint?   If a sprinter's arm swing and leg thrust (see figure 13.1) in any direction other than parallel to the direction of sprint, the forces that the sprinter applies to the earth in the direction of sprint are reduced. In reaction, the force that the earth applies against the sprinter is lessened as well. The result is that the sprinter doesn't run as fast as possible.

Why should a freestyle swimmer pull with the hands and forearms along a line parallel to the long axis of the body rather than emphasize an S-shaped out–in–up–down pattern of pull? Emphasizing an S-shaped out–in–up–down motion with the hands during the freestyle stroke, as shown in figure 13.10, is now thought to generate less propulsive force than pulling straight back against the water. A modified S-shaped motion still occurs during entry and exit of the swimmer's hand, but these actions occur more from body roll and the anatomy of the swimmer's body than from efforts to generate more propulsion. Pulling back against the water as far as possible parallel to the long axis of the body is now considered the correct technique. Under water, the arms flex at the elbows so that the swimmer's hands and forearms provide the major propulsive surfaces.

Why must athletes have their center of gravity positioned behind the jumping foot as they enter a high-jump takeoff or behind both feet as they prepare to jump to block or spike in volleyball? Positioning the takeoff foot ahead of the center of gravity gives the athlete more time to apply force with the jumping leg at takeoff (see figure 13.2). The athlete rocks forward, up, and then over the jumping foot. This large arc of movement gives the athlete time to drive down at the earth. The earth in reaction drives the athlete upward. The same principle applies to a volleyball spike, a volleyball block, a basketball layup, and a basketball block.

Why is it important for athletes to rotate the hips and thrust them ahead of the upper body during a golf drive, shot put, and discus or javelin throw? Rotating the hips ahead of the upper body and toward the direction of throw serves three purposes. First, it shifts the athlete's body mass in the proper direction (i.e., toward the direction in which the golf club, discus, shot, javelin, or baseball bat will be accelerated). This action extends the distance and time over which the athlete applies force. Second, the rotation of the hips acts as an important link in the sequential acceleration of the athlete's body segments. The movement of the athlete's legs and hips toward the direction of throw (or impact with the ball in golf or baseball) simulates swinging a whip handle ahead of the rest of the whip so that the tip of the whip cracks. Third, the rotation of the hips stretches the muscles of the abdomen and chest so that they pull the shoulders and throwing arm in slingshot fashion toward the direction of throw.

(Notice the weight shift and hip action in the javelin throw in figure 13.4 and in the golf drive in figure 11.4.)

Why should athletes extend the kicking leg when contacting the ball in a football punt? By extending the kicking leg, the athlete puts the part of the foot that contacts the ball farther from the axis of rotation (i.e., the hip joint). Because of this increase in radius, the kicking foot is moving faster than any other part of the leg when it contacts the ball. The flexion of the kicking leg before contact with the ball, together with its extension at impact, simulates a whiplash action (see figure 13.11).

AT A GLANCE
Analyzing Sport Skills

- One of the greatest challenges you'll face working in sport is watching the athlete perform and deciding which aspect of the skill needs correction (if any).

- All fundamental actions that an athlete makes within technique are founded on mechanical principles. In other words, technique is based on mechanical laws.

- After you have analyzed the performance, you need to communicate this information to the athlete. Technology can be an effective mechanism for providing feedback.

Why must athletes extend their bodies fully at takeoff in gymnastics and diving skills? Athletes who need to rotate quickly must apply an **eccentric thrust**, or an off-center force, at takeoff to initiate rotation. They must then pull the body inward from a fully extended position. The large reduction in rotary inertia caused by compacting the body mass around the axis of rotation is rewarded by a huge increase in the rate of spin (i.e., angular velocity).

All phases and all key elements in a skill are performed for specific mechanical purposes. If you know the mechanical reasons that they're performed as they are, you can confidently say to yourself, "OK, I understand what should occur in the technique of this skill, and I understand the mechanical principles behind the movements that the athlete must perform. I'm ready to watch my athlete and correct any errors that I find."

We have asked you to use elite performances as a model or reference point when working in

Rowers Use Hatchet Blades to Apply More Force

High-performance rowers use oars with huge blades that look like giant meat cleavers. Called hatchet blades, these oars are shorter from the oarlock to the blade than standard oars are. The mechanical principle behind this design is that for the same effort from the rower, the hatchet blade travels more slowly through the water but applies more force. Moving slower, the blade slips less in the water but propels the shell faster. Do these blades present any problems? According to many coaches, hatchet blades, although allowable within the rules of rowing at many levels, can cause stress injuries in the lower and upper limbs because the rowers have to pull against a stiffer and less mobile resistance.

Robert Cianflone/Getty Images

a sport. But don't make the mistake of trying to mold a young athlete in the exact image of a high-performance athlete. When you watch a series of top performances, be sure to study the basic technique that these top athletes use—nothing more. With your knowledge of mechanics, you'll see the purpose behind these actions. As your knowledge of sport mechanics improves, you'll learn to disregard some actions that a top-class athlete uses because they are personal idiosyncrasies of no mechanical value. Accept them as something that makes an individual athlete comfortable but disregard them as a necessity for good performance.

Remember that the actions that an elite athlete performs at high velocity over a great range of movement need to be modified to be appropriate for the maturity, strength, flexibility, and endurance of a young athlete. You cannot and must not expect a young, immature athlete or a novice of any age to assume the body positions or match the explosive actions of an experienced athlete. This development comes with regular training and good coaching.

SUMMARY

- Six steps are useful in analyzing a sport skill: (1) Determine the objectives of the skill; (2) note any special characteristics of the skill (these two steps highlight the objectives and conditions governing the performance of a skill); (3) study elite performances of the skill (this step requires careful analysis of elite performances of the skill you are coaching); (4) divide the skill into phases; (5) divide each phase into key elements (these last two steps show the importance of breaking down a skill into phases and key elements); and (6) understand the mechanical reasons that a key element is performed as it is (this step emphasizes the need to understand why the performance of the phases and key elements of a skill should be based on sound mechanical principles).

- After you have analyzed the performance, you need to communicate this information back to the athlete; technology is an effective mechanism for providing feedback.

- The rules of sport and the conditions that exist when sport skills are performed determine skill objectives. Most sport skills have more than one skill objective.

- Sport skills can be divided into different types based on the manner in which the athlete performs the skill and the conditions under which the skill is performed.

- A number of factors contribute to the final result or outcome; some of these subcomponents are fixed, and others can be changed.

- Nonrepetitive skills are also called discrete skills because they have a definite beginning and end. Nonrepetitive skills are frequently joined in a sequence.

- Repetitive skills have a cyclic, continuous nature; the movement pattern occurs again and again.

- Sport skills can be performed in predictable and unpredictable environments. Skills performed in a predictable environment are called closed skills. Skills performed in an unpredictable environment are called open skills.

- A phase in a sport is part of a connected group of movements that an athlete joins together in the performance of the total skill.

- Many skills can be divided into the following four phases: (1) preparatory movements (setup) and mental set, (2) windup (backswing), (3) force-producing movements, and (4) follow-through (recovery).

- Key elements are the finer, distinct actions that together make up a phase. Force-producing movements generally contain the most key elements.

- Analysis of sporting technique can be done in **real time**, and a more complex analysis can be conducted after the performance.

- Various forms of technology can provide a novel report on what is happening within a performance; examples are split-screen video imaging, player and object tracking, and complete sport analysis.

- An understanding of the mechanics of the key elements and phases of a sport skill is necessary to teach a technically correct performance of a sport skill.

KEY TERMS

closed skills

discrete skills

eccentric thrust

follow-through

force-producing movements

nonrepetitive skills

open skills

phase

real time

repetitive skills

skill objectives

windup

REFERENCES

Farrow, D., and S. Robertson. 2017. "Development of a Skill Acquisition Periodisation Framework for High-Performance Sport." *Sports Medicine* 47 (6): 1043-1054.

Fedorcik, G.G., R.M. Queen, A.N. Abbey, C.T. Moorman, and D.S. Ruch. 2012. "Differences in Wrist Mechanics During the Golf Swing Based on Golf Handicap." *Journal of Science and Medicine in Sport* 15 (3): 250-54.

Ross, J.A., C.J. Wilson, J.W.L. Keogh, K.W. Ho, and C. Lorenzen. 2015. "Snatch Trajectory of Elite Level Girevoy (Kettlebell) Sport Athletes and Its Implications to Strength and Conditioning Coaching." *International Journal of Sports Science and Coaching* 10 (2-3): 439-52.

Travassos, B., K. Davids, D. Araújo, and T.P. Esteves. 2013. "Performance Analysis in Team Sports: Advances From an Ecological Dynamics Approach." *International Journal of Performance Analysis in Sport* 13(1): 83-95.

Van Biesen, D., K. McCulloch, and Y.C. Vanlandewijck. 2017. "Comparison of Shot-Put Release Parameters and Consistency in Performance Between Elite Throwers With and Without Intellectual Impairment." *International Journal of Sports Science and Coaching*: 1747954117707483.

Young, W., S. Grace, and S. Talpey. 2014. "Association Between Leg Power and Sprinting Technique With 20-m Sprint Performance in Elite Junior Australian Football Players." *International Journal of Sports Science and Coaching* 9 (5): 1153-60.

 Visit the web resource for review questions and practical activities for the chapter.

12

Identifying and Correcting Errors in Sport Skills

Hero Images/Getty Images

When you finish reading this chapter, you should be able to explain

- how to observe the performance of a skill;
- how to analyze each phase of a skill and the key elements within each phase;
- how to make use of your knowledge of sport mechanics in your analysis;
- how to determine the order in which to correct errors;
- how to select the appropriate coaching methods for correcting errors; and
- how to determine the correct technique for your athlete.

In chapter 11 you went through the first stage in getting ready to correct errors in an athlete's performance. You saw how to break a skill into phases and key elements, and you understood how sound mechanical principles form the foundation of good technique. This chapter gives you advice about observing an athlete's performance and using your knowledge of mechanics to pick out errors that need to be corrected. We have laid out what you must do in a series of five steps:

Step 1: Observe the complete skill.

Step 2: Analyze each phase and its key elements.

Step 3: Use your knowledge of sport mechanics in your analysis.

Step 4: Select errors to be corrected.

Step 5: Decide on appropriate methods for the correction of errors.

Step 1: Observe the Complete Skill

Planning how you intend to observe a skill is a good idea. The preferred approach is to watch the whole skill several times and then home in on its **phases** and **key elements**. From this process, you can decide what to look at and where to stand. Watch from the left and right, from the front and rear. In this way you can cross-reference and double-check the information you gather. Characteristics of the performance that are hidden from one point of view will be revealed from another.

Ensure Safety When Observing

Before you begin to observe a skill, you must ensure safety for the coach and the athlete. Note that viewing from the front is not recommended for throwing events in track and field or for sports such as golf, in which the velocity of the ball is exceptional. Unless you have a specially designed protective screen, as used in baseball, be satisfied with viewpoints from the side and the rear.

Choosing a Setting for the Observation

When choosing a setting for the **observation**, try to avoid conditions that distract you and the athlete. Physical education classes, recreational settings, or training sessions within the sport can disturb your concentration and that of the athlete because too many other activities are going on. Any movement in the background can disrupt your attention from the details you want to analyze. If you are instructing a group, you cannot, and should not, pay attention to one athlete for long. Other athletes need your supervision and encouragement. The best setting is one in which no distractions at all are present. The athlete can concentrate on performing, and you can focus on observing and analyzing.

Positioning During the Observation

With many skills, you'll pick up much worthwhile information when you observe from in front of the athlete. But be particularly careful in this situation. You may center your concentration on the athlete's movements and not on what happens afterward.

With rotational skills such as discus, shot put, and hammer throw, you should be well back behind a safety cage officially approved for the event. This viewing position is highly recommended for the discus and hammer throw. In the hammer throw, the 7.2 kg (16 lb) ball travels at phenomenal velocity, and in the hands of a novice it may not fly in the required direction.

- If no safety cage is available, stand well back to the left rear (as you view the athlete from the rear) for hammer and discus throwers who rotate counterclockwise across the ring.
- In the shot-put event (which normally does not use a cage), stand well back to the left rear of right-handed throwers and to the

right rear of left-handed throwers. Be sure to stand well to the rear of the throwing ring if the athlete is learning the rotary shot-put technique.

If you are marking distances while you assess the athlete's performance, be aware that an implement in flight is extremely deceptive. Javelins viewed head-on have a habit of momentarily seeming to disappear from sight, and wind can dramatically alter flight paths. You must also allow for the distances over which implements skid and bounce. A discus skidding across wet grass is extremely dangerous!

In addition, skills that involve height and flight (e.g., gymnastics vaulting, ski jumping, and pole vaulting) can be more demanding to observe than skills that contain much less movement (e.g., archery or power lifting).

A gymnastics vault includes a long and fast approach, a takeoff, flight onto and off the horse, and finally a landing. These phases of the skill occur at high speed and cover considerable distance and height. To observe all aspects of the action critically, observe from various positions:

- Stand at right angles to the board about 4.6 m (15 ft) from the flow of the skill.
- Variations can include positioning yourself to the rear of the approach.
- You can also stand beyond the landing pads so that the athlete runs toward you. In this way you'll get several viewpoints of the takeoff, flight, and hand positions on the horse, as shown in figure 12.1.
- This observational technique also works well for track events, as shown in figure 12.2, in which a track coach observes from various viewpoints while the athlete practices hurdle clearances.

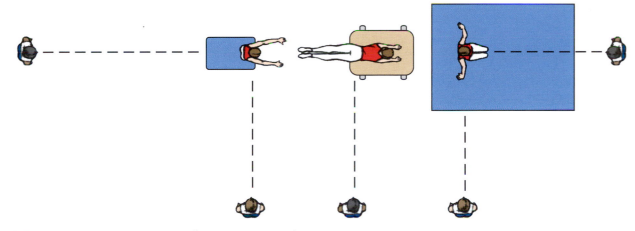

FIGURE 12.1 Various viewpoints for assessing a vault in gymnastics.

Adapted from J.G. Hay and J.G. Reid, *Anatomy, Mechanics, and Human Motion*, 2nd ed. (Englewood Cliffs, NJ: Prentice-Hall, 1988), 258.

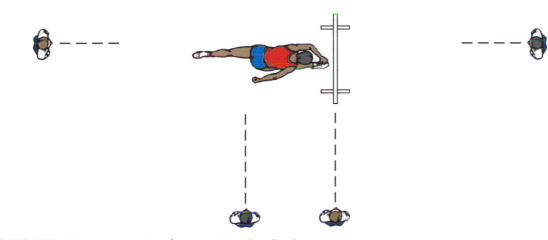

FIGURE 12.2 Various viewpoints for assessing a hurdle clearance.

- Get closer when skills cover less distance and height and when you are focusing on particular phases and elements of the skill.

The ultrahigh quality video clarity from common video recorders (such as mobile phones) and the video images from television broadcasts provide excellent slow-motion coverage of athletes viewed from above. You're probably familiar with the dramatic replays of hand changes on the high bar, swings to a handstand on the rings, and the incredible rotary skills of gymnasts on the pommel horse. In swimming, cameras on tracks at the sides and bottom of the pool give superb coverage of swimming strokes. This additional visual information will help you tremendously in assessing the athlete's performance.

Also note that if you are a novice working in the sport, you may find it difficult to observe a performance critically when you are also involved in spotting. Your attention tends to focus on where you should give support (and perhaps on protecting yourself from the flailing arms and legs!) rather than on whether certain movements are performed correctly. In gymnastics, dividing your attention can be a risky practice. Experienced coaches and sport scientists can carry out both jobs at once, but they must concentrate on their spotting when more complex skills are attempted. If you are starting out in a sport that has a high level of risk, play it safe and use competent spotters if you want to be free to observe. If spotters are not available, have an onlooker video the performer while you give the necessary assistance. Afterward, analyze and discuss the performance with the athlete and sport staff.

Observe an Athlete's Performance of the Skill

When you are ready to observe the athlete, have the athlete warm up and perform the complete skill several times so that you get a good overall impression. Don't concentrate on specific phases, even though a poor windup or poor force-producing phase will obviously catch your eye. Try to get a feel for the athlete's rhythm, flow, and general body positions from the start of the skill to the finish. Your main objective at this stage is to get an overall impression of the athlete's performance.

When you observe a complete skill this first time, the athlete should perform the skill at normal speed. The reason for this recommendation is that skills performed at unnaturally slow speeds are dramatically different from those that occur at normal speed.

- Timing, coordination, and the feel for the skill are different. Slow-speed performances serve little purpose when you are looking for errors to correct. They give you a false picture of what is occurring.
- Reducing speed, however, is helpful when you teach new movement patterns.
- When the fundamentals have been taught and learned, the speed of movement can be increased.

The number of times that the athlete performs the skill for observation depends on the physical demands of the skill. Skills that take considerable time, concentration, and effort for each repetition, such as diving and ski jumping, are by necessity viewed fewer times than a place kick, a volleyball serve, a pass in soccer, or the repetitive paddling actions of a kayaker. Nevertheless, you need to view the performance enough times that the athlete's pattern of movements becomes apparent. More than one training session may be required to develop an accurate impression of the athlete's abilities in skills such as diving and ski jumping.

As you observe, expect that a beginner's performance will change dramatically from one repetition of the skill to the next and that a beginner will tire more quickly than an experienced athlete will. Novices may make gross errors in which they miss several key elements in a phase, or even a whole phase of a skill (Buszard et al. 2016). During your general observation you'll notice that their foot positions are incorrect at one moment and correct the next. The beginner may not use the large muscles of the body or shift the body weight in the correct direction. You may even think, after completing the observation, that the best course of action is to rebuild the skill completely. With novices you must accept this situation as part of working in the sport with beginners!

In comparison with beginners, high-performance athletes make fewer apparent errors. You'll see errors when you analyze slow-motion video of the skill, or you'll catch errors when you concentrate on specific elements in the skill. Perhaps you'll discover that the athlete's line of vision is incorrect or that his head is in the wrong position, upsetting his balance. You may discover that an athlete's overall performance is good except that the wrist action at the end of a pitch, throw, or hit is not as it should be. Unfortunately, you may also have to struggle through many training sessions to get a high-performance athlete to eliminate these seemingly minor errors. The reason for this difficulty is that a top athlete has probably been performing the skill in the same way for years, thereby ingraining the incorrect action. Coaching a young novice is entirely different. Every coaching session can be a giant leap forward! To your delight you'll find that the technique of most novice athletes is like clay that you can mold. Each training session can result in a massive change in the quality of their performance. For that reason, many professionals involved in sport find great pleasure in working with novice athletes.

- Also note that when you observe, you should not distract the athlete by continually offering instruction. Watch without making any comments other than an occasional encouraging remark after the skill is completed.

- Try to keep the athlete relaxed and enjoying having you as an enthusiastic and knowledgeable spectator. The athlete should not struggle to impress you, become discouraged, or become so casual that she loses concentration. You want an accurate impression of her abilities, not a performance altered by tension or insufficient concentration.

- Above all, don't start listing aloud all the actions that the athlete is doing wrong! This commentary serves no purpose and destroys morale. You don't want the athlete to become tense or stressed in any way. Your job is to get a true impression of how he performs.

- Also, while you observe, make a mental note if you think that the athlete is lacking in strength, flexibility, or endurance. But remember that she cannot change these characteristics during one training session, just as an athlete cannot gain or lose weight on command. Take these factors into account by modifying your demands when you start correcting errors. In other training sessions you can get the athlete to work on improving those areas.

Look for Other Clues in the Observation

Part of your observation technique will be to look beyond the athlete for clues about the performance. The flight path, rebound, and roll of balls result from the movements and actions that the athlete uses in the skill. Skate marks on ice, ski patterns on snow, and footprints on approaches, takeoffs, and landings are all clues to what's going on in the skill.

APPLICATION TO SPORT

Innovative Technique Produces High-Performance Skiing

At the 1976 Winter Games in Innsbruck, downhill skier Franz Klammer thrilled the world with his hair-raising gold medal performance. Klammer's rocketing do-or-die performance is still considered one of the most exciting and brilliant downhill runs in the history of modern ski racing. In remembering that run many years ago, Klammer said he was helped immensely by superior coaching and equipment. He also said that a large part of his success came from the innovative way in which he carved his turns. At the time, ski racers used to skid through their turns on the flats of their skis. Klammer started his turn with the ski on its edge. With less skidding at the beginning of the turn, Klammer came out of his turns with more speed than his rivals. After Innsbruck, in a World Cup career that spanned nearly 15 years, Klammer amassed 25 downhill victories, the most for any downhill ski racer. In 1975 and 1976 he won 13 World Cup downhills in a row, a record that remains unbroken.

GEORGES BENDRIHEM/AFP/Getty Images

- Use your senses (your ears and eyes) when you look for clues! The rhythm of footfalls during an approach or during the repetitive bounding of a triple jump is an indication of stride length and stride cadence.

- The overemphasized thud of an athlete's feet during throwing events is a sure sign of poor balance and weight distribution. (It's also a sure sign that the hop or the step in the triple jump is too large.)

- The noise of bat and club on ball helps distinguish a direct impact from a sliced impact.

- In volleyball, a slapping noise is a giveaway of a carried ball or some other incorrect contact.

- Almost every sport will give you visual and auditory signals that you'll be able to associate with good or bad performance. Use every source of information. Don't limit yourself in any way.

AT A GLANCE

Observation of Sport Skills

- Before you begin to observe a skill, you must ensure safety for the coach and the athlete.

- Make observations from several viewpoints to observe all aspects of the action critically. Observe from positions that are at right angles to the flow of the skill and if possible from the front, rear, above, and below.

- When you observe a complete skill this first time, the athlete should perform the skill at normal speed. Skills performed at unnaturally slow speeds are dramatically different from those that occur at normal speed.

Step 2: Analyze Each Phase and Its Key Elements

After you have watched the complete skill several times, you are ready to concentrate on individual phases and their key elements, as discussed in chapter 11 and shown in figure 11.5. You can approach this task in two ways: observe the result and work back or observe each phase of the skill in **sequence**.

Observe the Result and Work Back

One method commonly used in correcting errors in a sport with a competent athlete is to start with the end product and work back from there. Here's an example: A player in rugby is trying to spiral the ball for distance. The player puts enough force into the kick, but there's no spiral. Therefore, you concentrate on the action of the foot as it contacts the ball (Sinclair et al. 2016).

- Is the ball fed onto the player's foot correctly?

- Is the foot drawn across the long axis of the ball to produce the torque necessary to generate a spiral?

- On the other hand, if the ball's spiral is satisfactory but distance is lacking, then shift your attention to other phases and key elements in the skill.

- Does the player shift the body adequately into the kick?

- Are the lower leg and kicking foot allowed to swing freely, or is the kicker tightening the leg muscles and eliminating any chance of producing a whiplash action?

- Is flexibility a problem? Inflexibility will restrict the range that the kicking leg swings through and reduce the force applied to the ball.

In throwing, kicking, and striking skills, checking the result gives you a wealth of information. For example, you might have a big, powerful shot-putter who ought to throw the shot a long distance, but the force behind the shot is inadequate. You know that the athlete should be throwing 1.5 m (5 ft) farther. So you concentrate on the throwing stance that the athlete assumes after completing the glide across the ring. When you examine the throwing stance, ask yourself these questions:

- Is the athlete's body angled correctly?

- Are the shoulders still facing the rear of the ring when the glide is complete?

- Is the foot placement correct?

- Is the athlete rotating the hips toward the direction of throw, and are the massive muscles of the legs, seat, and back used before the chest, arm, and fingers? (Figure 12.3 shows the actions that you should be looking for in shot put.)

After critically examining the throwing stance, you may decide that the problem lies in a poor glide across the ring. Well-performed standing throws confirm your suspicion. The athlete's glide across the ring is ruining the remaining part of the

FIGURE 12.3 Excellent technique using the glide method in the shot put.

throw! With the problem diagnosed, you and the athlete can work on correcting the errors in this phase of the throw.

Observe Each Phase of the Skill in Sequence

Another method of observation commonly used in sport is to start by critically observing the first phase in the skill and then to progress to the second, third, and so on. The first phase contains preliminary movements and the athlete's establishment of a mental set. In the first phase, look at such elements as the athlete's stance and weight distribution. Take note of his head position, his line of vision, and the way he concentrates for the actions that will follow.

In the second phase, when the athlete winds up or performs a backswing, examine the weight transference from one foot to the other (figure 11.4a and 11.4b illustrate proper body position, weight shift, and backswing in a golf drive). Check the position of the implement and the athlete's body at the end of the windup. Make a mental note if the athlete appears stiff and needs to improve flexibility.

When you examine the force-producing phase, remember that in many skills this phase is made up of several distinct sections:

- An approach and a takeoff in jumps and vaults
- An approach, glide, spin, and throw in throwing events
- An approach, hurdle step, board flexion, and takeoff in springboard dives

In freestyle swimming, the force-producing phase can be the catch of the water at hand entry followed by a long pull and push from the hand and forearm. Break these complicated force-producing phases into key elements and concentrate on each key element in sequence.

In most skills, the follow-through is the least important of all phases. Your athlete has applied force, and the follow-through safely dissipates their momentum and kinetic energy. But be sure to observe what happens during the follow-through and, of course, what happens to the implement and the athlete immediately afterward. The athlete's actions and the flight of the implement are clues to what happened earlier. Check the athlete's arm and hand actions during the follow-through on a jump shot in basketball or a volleyball spike. In these skills, a follow-through can indicate the amount and direction of force and spin that were applied to the ball.

In some sport skills, insufficient control during the force-producing phase produces a follow-through that violates the rules of the sport. A field hockey player may swing the stick too high, or a volleyball player may hit the net after spiking or blocking. So don't disregard the follow-through; consider it an important phase that gives you clues about what happened during the windup and force-producing phases that occurred earlier.

In cyclic, repetitive skills such as swimming, remember that the follow-through is a recovery action that sets up the athlete for another force-producing phase. Check that these recovery actions are mechanically efficient and not wasting the athlete's energy. A misdirected arm recovery in freestyle, in which the swimmer's arm swings across the midline of the body, produces poor body alignment, generates excessive form drag, and affects the efficiency of the force-producing phase that follows (see figure 12.4). Cyclists talk of the need to spin the pedals, meaning that proper pedaling technique is a rotary motion, not just a push downward with a rest on the way up. Check

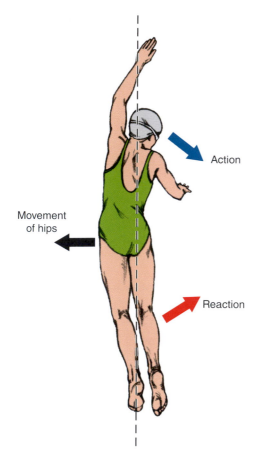

Action

Movement
of hips

Reaction

FIGURE 12.4 Overreaching and crossing the midline at arm entry causes the hips to react and move in the opposing direction.

Adapted from E. Maglischo, *Swimming Fastest* (Champaign, IL: Human Kinetics, 2003), 53.

that the athlete is pulling up and around with one leg while pushing down and around with the other. Proper pedaling technique is the ultimate in cyclic, repetitive action.

Step 3: Use Your Knowledge of Sport Mechanics in Your Analysis

As you observe each successive phase and its key elements (from preliminary movements and mental set to follow-through or recovery), you must put your knowledge of sport mechanics to work. In particular, concentrate on how the athlete applies muscular force to produce a desired action in a skill. You must carefully assess the mechanical efficiency of the athlete's actions and the way that she competes against gravity, friction, drag, air resistance, and the forces generated by opponents,

whatever the opposition might be. In this way you can pick out technical (i.e., mechanical) errors that the athlete commits. What should you be looking for as you examine the elements in each phase?

To assist in this process, having a record of the performance, such as a video, is often useful so that you can view the performance frame by frame to assist in the analysis. From this, you can piece together the information about what is happening in the movement and, more important, use this knowledge to enhance the athlete's performance. To work through this process, here is a series of important questions you can ask yourself:

Does the athlete have optimal stability when applying or receiving force? A wide base and correctly positioned center of gravity are essential for applying and receiving force. Check the position of your athlete's center of gravity and the way he sets up a base of support. Ask yourself the following subquestions:

- Is the base of support extended in the direction it should be?
- Is the base too narrow or too wide?
- Is the athlete standing too erect instead of squatting down?
- Is the center of gravity too close to the edge of the base when it should be centralized?

If the athlete stumbles or gets thrown one way as the implement goes the other way, or if she is too easily knocked off balance by an opponent, you'll know that an error has occurred in this area and you'll need to check the mechanical principles associated with balance and stability. Remember that stability has to do with one turning effect (i.e., torque) battling another. To remain stable, your athlete may have to reposition her feet and center of gravity to apply more leverage and more torque (Zemková 2014).

In many skills, the athlete must be able to move quickly and react in an instant. When he is receiving a serve, playing in goal, or reacting to the moves of an opponent, the objective is not maximum stability but rather a level of stability that allows him to move in a flash in any direction. We discussed these principles in detail in chapter 8.

In particular, be sure to check the size and alignment of the athlete's base of support and the position of his center of gravity during the force-producing phase of a skill. An inadequate base not only makes the athlete unstable but, equally important, also reduces the distance and time over which he can apply force.

Is the athlete using all the muscles that can contribute to the skill? Athletes produce inferior performances when they do not apply force with all the muscles they can and should use in a skill. This state of affairs may seem strange! After all, why not use the leg muscles, or any other muscle group, if they can contribute to the performance? If the performance of a skill requires the muscles of the legs, trunk, chest, and arms and the athlete uses only the muscles of the chest and arms, then the total force put into the skill will be below the optimal level. How can you tell if your athlete is using all the necessary muscles? Usually, the answer is easy because in dynamic skills, muscle contractions produce actions. If a limb segment or some other part of the body moves, you know that muscles are contracting. Here's an example of what we mean:

- When a child throws a ball for distance for the first time, the youngster frequently stands still with the feet close together and then throws with the arm alone. The child doesn't use a wide throwing stance.
- The child doesn't take back the throwing arm or rotate the shoulders away from the direction of throw.
- In the force-producing phase of the throw, the muscles in the legs, trunk, and chest do not contribute.
- This error doesn't occur only among children. You'll see it happening among adults as well.

APPLICATION TO SPORT

What Is the Correct Technique for a World-Record Performance?

A common dilemma for an athlete and sport professional involves determining the correct technique for a perfect performance. Should the technique be based on the current world-record holder, or should something else be considered? The trap for young players, coaches, and sport scientists is to adapt or modify the technique to mimic the current world champion. Although we should observe the new outstanding performance, we fundamentally need to break down the skills into the mechanical components. From this information we need to determine the appropriate technique.

Clive Rose/Getty Images

At the 2016 Rio de Janeiro Games, Michael Phelps came out of retirement at the age of 31 to win his 19th gold medal from five Olympic Games. Many of his races were won in world-record time. Detailed analysis showed that Phelps was similar in height and weight to the other swimmers, but an anthropometric analysis demonstrated that his upper body was slightly longer than his lower limbs. Was this his secret weapon? Do we need our athletes to adopt this "technique"? If we look back to previous Olympics Games, in 2004 at Athens and in 2000 at Sydney, Ian Thorpe won five gold medals. He was similar in height and weight to the other swimmers but had larger feet. Was that his secret weapon? Do we need our athletes to adopt this "technique"?

Because changing these physical attributes is not possible, we want to ask whether any other factors contributed to Phelps' world-record performances. A key difference with Michael Phelps was his race strategy; his stroke rate and stroke length were different from those of former world-record holders such as Ian Thorpe and Pieter van den Hoogenband. So what is the correct technique? That is the question for coaches. The answer is that it all depends on the attributes of the individual athlete.

Top-flight athletes always aim to have all the required muscle groups contribute to the skill. High-performance rowers make sure that the muscles of the legs, back, shoulders, and arms play their part in the stroke. Top speed skaters make sure that their leg muscles contribute optimally in powering them along the ice. The muscles working their arms and shoulders make their own contribution in counterbalancing the actions of the legs. Imagine how poor the performance would be if a speed skater failed to use the quadriceps muscles adequately to extend the legs or skated with the arms hanging straight down instead of forcefully swinging them back and forth! The same principle applies to all skills. Make sure that all the athlete's muscles are contributing by moving the body segments they are responsible for moving. Think of the athlete's muscles as a tug-of-war team. No member of the team should rest on the rope without pulling.

Is the athlete applying force with the muscles in the correct sequence? If a world champion weightlifter performs a clean and jerk and you critically examine the key elements of the clean (in which the athlete pulls the bar up to the chest), you'll see that the muscles of the legs, back, shoulders, and arms are contracted at about the same time. The extension of the legs is closely linked to an extension of the back and a strong upward pull with the arms (McKean and Burkett 2010).

On the other hand, if you examine the key elements in the force-producing phase of a pitcher's fastball, you'll see a well-defined sequence of actions, starting from the big muscles that accelerate the athlete's body and the large, more massive body segments and finishing with the high-speed movement of smaller, less massive body segments (i.e., the throwing arm and hand). All great pitchers step forward as they draw the throwing arm back and rotate the shoulders away from the batter. When they have stepped out into the pitching stance, their bodies rotate toward the hitter in a whip-like sequence that starts from the legs, shifts to the hips and then to the chest, and ends with a tremendous acceleration of the throwing arm. The pitcher's body acts like the handle of a whip that is being cracked. The hand gripping the baseball is the tip of the whip. A volleyball spike or a tennis serve uses a similar whip-like sequence of actions.

This comparison between the actions used in a clean and jerk and a baseball pitch demonstrates opposite extremes in the sequence in which an athlete's muscle contractions occur.

- When you examine each phase in the athlete's performance, check that the movement of his limb segments occurs in the correct sequence.
- If it does, then you know that muscle contractions are occurring in the correct sequence as well.
- A common fault for many athletes in throwing and hitting skills is to use the small muscles of the shoulders and throwing arm long before the big muscles of the legs, back, and trunk have done their job.
- The result is that the big muscles never get the heavier parts of the body moving ahead of those that are lighter. Cracking a whip is impossible if you don't accelerate the whip handle first.

AT A GLANCE

Applying Sport Mechanics to Correct Sport Skills

- The assessment of technique evaluates the mechanical efficiency of the athlete's actions and the way they compete against gravity, friction, drag, air resistance, and the forces generated by opponents, whatever the opposition might be.

- In some sport skills, insufficient control during the force-producing phase produces a follow-through that violates the rules of the sport.

- In cyclic, repetitive skills, the follow-through is a recovery action that sets up the athlete for another force-producing phase.

- Overall, does the athlete apply force with the muscles in the correct sequence?

Is the athlete applying the right amount of muscular force over the appropriate time and distance? You'll recognize that this question refers to our old friend impulse, which we discussed in chapter 5. Remember that impulse has to do not only with the amount of force that the athlete uses but also with the time over which the athlete applies force.

If the athlete uses the right amount of force for the right amount of time, the limbs move at the required speed through the required range of movement. When this occurs and all muscle contractions are sequenced correctly, you'll see movements that are fluid, smooth, rhythmic, graceful, and well coordinated. When an athlete applies force indiscriminately and haphazardly (and this is what novices do!), you'll see actions that are jerky and awkward. What you are seeing is the difference between a polished and well-practiced technique and one that is not.

Practice helps athletes establish how much force each muscle involved in a skill must exert. When they are learning, many athletes apply too much or too little force at the wrong time, so their technique looks jerky and awkward. You can help the athlete correct this situation by giving rhythmic cues that provide an idea of the speed at which she should perform the actions. You can also provide tips that translate your sport mechanics knowledge into a language that the athlete can relate to, such as, "Step out long and low, and as soon as your foot hits the deck, thrust your hips toward the direction you are throwing" or "Stretch up at takeoff and swing your arms upward as fast as you can."

Keep in mind that what is correct in a mechanical sense is not always possible in an anatomical sense. In other words, mechanical principles must fit with the design of the athlete's body. For example, athletes do not apply maximum force over the longest possible time even in hitting skills that require maximum velocity at impact or in jumping skills that require maximum velocity at takeoff. In these skills, good technique is characterized by limbs that are slightly flexed at the start of the skill and fully extended from maximal muscular contraction when impact and takeoff occur. Look for this action when you examine the final elements of the force-producing phases in golf driving, as shown in figure 11.4, baseball batting, tennis serving, and track and field throwing.

In skills that require accuracy, such as a volley in tennis or a drop shot in squash, look for controlled force applied over a specific range. Too much force or too great a range of movement defeats the purpose of the skill. When too much force or too great a range of movement occurs, the volley puts the ball out of the court. Likewise a drop shot in squash rebounds too high from the front wall, making the "get" easy for the opponent.

A word of advice in relation to force and the period over which force is applied: No one expects young athletes to be able to produce the same force as adults do. Likewise, you cannot expect novices to assume the same body positions and apply force over the time and distances that high-performance athletes use. Make allowances when you watch a young athlete perform. Less force applied over a limited range of movement is not necessarily an error but can be a stage in the developmental process. More force and a greater range of movement will come with increased strength, flexibility, endur-

ance, and coordination—all of which are carefully molded by your good **feedback** when identifying and correcting errors in sport.

Is the athlete applying force in the correct direction?   This principle may seem hardly worth mentioning, yet you should look for this factor, particularly in the force-producing phase of the skill. High-performance sprinters drive down and back with each leg thrust so that they travel forward at the greatest possible velocity toward the finish line. The direction of each leg thrust gives these superb athletes the exact amount of vertical and horizontal thrust required by each sprinting stride. The result is optimal forward propulsion. You'd have no trouble deciding that something was terribly wrong with a sprinter's technique if his leg thrust was directed out to the side or if his arms swung sideways across his body rather than forward and backward! Errors like these in the force-producing phase of sprinting indicate that the athlete is wasting force and not applying it in the correct direction.

In your analysis of the performance of sport skills, you'll notice that inexperienced athletes apply force in many directions. Much of their muscular effort makes no contribution to their performance.

- Look also for inexperienced athletes to thrust and push in the correct direction with one part of their bodies and in an incorrect direction with another part. You see this often with novice downhill skiers.

- Young shot-putters often complain that putting the shot bends their fingers backward. Careful examination of the arm action in their throws indicates that they are not pushing directly behind the center of gravity of the shot. The thrust is in some other direction, and the result is that the shot bends their fingers back.

- Likewise, a hammer thrower or discus thrower who spins on the spot instead of traveling across the ring or who falls sideways out of the ring when releasing the implement is obviously misdirecting her force!

- Poorly directed force produces inadequate rotation at takeoff for gymnasts, divers, and figure skaters; mis-hits in sports that use clubs, bats, and rackets; mis-kicks in sports that use kicking skills; and poor propulsion in swimming skills.

Is the athlete correctly applying torque and momentum transfer?   Many sport skills require an athlete to generate and control rotation. Rotation is applied to the athlete's body, an opponent, a ball, or an implement such as a discus. To initiate rotation, the athlete must apply the turning effect of torque. The more spin that is required, the more torque the athlete has to apply. In your analysis, check how much force the athlete is generating and the distance at which this force is applied relative to the axis of rotation.

- In judo, look at the position of the axis of rotation that the athlete sets up for a hip throw and where he applies force to the opponent.

- In swimming, does the torque generated by the athlete transfer into the sport performance (Dingley et al. 2015)?

- Is your athlete strong enough to apply tremendous force? If not, is there any way of increasing the force arm (i.e., the distance from the axis to where force is applied)? The larger this distance is, the less effort the athlete has to apply.

In gymnastics, diving, and figure skating, the number of rotations that the athlete performs depends on how much torque she generates and how much momentum transfer she uses at takeoff. Momentum comes largely from arm and leg actions that are transferred at takeoff to the body as a whole. A skater who performs a double spin when intending to perform a triple may say, "I don't think I got enough spin." After you've analyzed the skater's takeoff, however, you may disagree. Check the actions of the arms and the free leg to see whether they contribute adequately to rotation. You might decide that sufficient torque was applied at takeoff, and in fact that it was overemphasized. Your analysis may indicate instead that upward thrust at takeoff was insufficient and that the arms and free leg were not swung vigorously enough to provide any transfer of momentum in an upward direction to the skater's body.

Is the athlete decreasing rotary resistance to spin faster and increasing rotary resistance to spin slower?   If a skill requires the athlete to spin faster, turn quicker, or swing the limbs at high speed, he must decrease his rotary resistance (i.e., rotary inertia) by pulling his body in toward the axis of rotation. The requirements of the skill determine the tightness of this position.

- Arms flexed at the elbows help produce a fast and efficient arm swing for sprinters and speed skaters.

- A tight tuck for gymnasts and a compressed body position for figure skaters when they spin around their long axis help produce the required number of rotations.

- Extended body positions oppose a fast spin and slow down rotation. Is your athlete not tucking tight enough because of insufficient flexibility or lack of muscular strength, or is the problem inadequate knowledge of the correct timing in the skill?

- Your careful analysis of each phase and its key elements can pinpoint the source of the problem.

Keep in mind that when the athlete rotates and extends the arms, the body slows but the arms and hands travel faster. Put a racket or a bat in the athlete's hand, and the head of the implement travels fastest of all.

Many skills require a combination of hip and shoulder rotation coupled with full extension of the arm (particularly at impact or release). For example, a tennis serve and a golf drive require full arm extension when the racket or club hits the ball. A discus thrower must release the discus with the implement as far from the body as possible. In your analysis, look for extension at release and impact and for some flexion and tighter body positions earlier in the skill.

Step 4: Select Errors to Be Corrected

After you've analyzed each phase and its key elements, you have the task of deciding what sequence to follow in correcting errors. Like any enthusiastic professional working in the sport, you'd like to correct every error in the first training session. The difficulty you'll face is that inexperienced athletes commit numerous errors, both major and minor. What is a major error, and what is a minor error?

- A major error is the absence or poor performance of any item discussed under step 3 in this chapter. Errors that destroy the athlete's stability or the optimal use of muscular force are major errors.

- A minor error is an action that only partially detracts from the performance of the skill. Examples of minor errors are a backswing in golf that needs to travel a few degrees farther back, a throwing stance that needs to extend a little farther, and an arm swing at takeoff in a jumping event that needs to be more vigorous.

- To a high-performance athlete, minor errors of this nature make the difference between a good performance and a world record. When working with a beginner, these errors can be placed on the back burner while other more important errors are corrected.

As shown in figure 11.5 in chapter 11, some factors in the performance can be modified or corrected, such as takeoff angle or body position. Others factors are more permanent, such as limb length or the dimensions of the sporting object (like the mass of the discus). The underlying rule when determining which errors to work on is to focus on the ones that can be modified.

The next step is to start with the fundamental components, or the core of the activity. A modification in this area will naturally lead to subsequent changes in other areas of the skill. Examples of key features are the following:

- Stance (a base of support)
- Grip (either on an object or the ground)
- Initial movement (for stability or for movement in the desired direction)

A skills checklist, as shown in table 11.1, can guide the next set of errors to focus on.

The simple method to follow when selecting errors is to forget those that are minor and pick out those that are major. When you've picked out the major errors, select the one that has the most adverse effect on the skill and work on that error first.

If you still have trouble deciding which error to choose, home in on major errors in the athlete's stance and body position—particularly in the preliminary stance and in the force-producing phase. Get the preliminary stance straightened out and then shift to the force-producing phase. You'll want to do this because an athlete cannot apply force correctly unless stance and body position are correct. In throwing, hitting, and striking skills and in contact sports, poor position and lack of balance destroy everything! An incorrect stance will ruin a golf stroke, and an incorrect body position sets up a wrestler for a countermove. In swimming, a body position in which the athlete is not horizontal in the water creates tremendous form drag. Keeping the head down and improving the leg kick can correct this. It's amazing how much faster a swimmer can travel when drooping legs are raised into a horizontal position.

Step 5: Decide on Appropriate Methods for the Correction of Errors

The final step delivers your application of sport mechanics and uses the traditional methods of teaching and coaching sport skills. Errors in skill performances vary in complexity. At their most complex, errors can be mistimed sequences of high-speed arm actions occurring in flight as a diver combines a somersault with a twist. At their simplest, errors may involve a young novice putting the wrong leg forward when throwing a softball. If you are working with a diver, you have the options of discussion, video analysis, and demonstration of the correct arm actions from the side of the pool. If you are lucky enough to coach at a pool with high-tech equipment, with the press of a button you can foam up the water with air bubbles so that your diver can work on the arm actions knowing that she is not going to get hurt if she fails. You might also decide that the correct arm actions need to be reinforced repeatedly on the trampoline with the diver in a spotting belt. But after the diver is in flight you cannot provide hands-on help and the athlete has no chance of slowing the skill in any way.

You can use the same approach when you work on improving a basketball layup or a start in swimming. Obviously, teaching a youngster to throw a softball is easier. You can translate your sport mechanics knowledge and demonstrate by saying, "Put your left leg forward and turn your right shoulder to the rear as you take your arm back." You can even move the youngster's limbs into the correct position. Doing this is impossible with a diver in flight or a swimmer leaving the blocks. Here are four suggested steps in correcting errors:

- First, you must ensure safety in coaching high-risk skills.
- Second, you must take steps in error correction.
- Third, you must communicate ideas for correcting errors.
- Fourth, you should use outside sources to aid in the correction.

Ensure Safety When Working in High-Risk Skills

The first step in choosing the appropriate methods for correction of errors is to ensure the safety of all involved. As mentioned earlier, in many skills the athlete cannot pause halfway through to rethink movements. These skills usually involve flight and often have a high element of risk. A back somersault in gymnastics floor exercises is such a skill. With skills of this type we recommend the following sequence:

1. *Maximize safety with spotters.* This precaution can include the use of overhead spotting rigs, safety belts, crash pads, pits filled with foam rubber, or other specialized equipment that fully protects the performer. In this way the athlete can perform the required actions with confidence and without danger.

2. *Begin with the accomplished fundamental skill before moving to complex tasks.* With highly complex skills, go back to a known skill that contains elements of the movement patterns you wish to correct. Use this skill to reinforce the correct actions.

3. *Progressively remove spotters.* As the skill is learned, the spotters and other specialized equipment can be carefully removed.

Take Steps in Error Correction

Because sport skills vary so much and errors are so diverse, no single method works for the correction of all errors. But we can provide a step-by-step sequence that will help in most situations you'll encounter.

1. Separate the phase that contains the error from the rest of the skill (if possible). Treat this phase and its key elements as a skill in itself.

2. Break the phase and its elements into smaller parts. For example, if the error is poor synchronization of footwork and arm actions, consider teaching the footwork first. Then teach the arm actions. Later add the two together. Use oral counts and rhythmic cues to assist the athlete.

3. Design a practice or a specific activity that is useful for teaching the correct movements. This practice should be easy to perform and, if possible, novel and interesting to the performer. Above all, be creative and flexible in your approach. If the practice activity you

have designed doesn't help correct the error, change it. What works well for one athlete may not work so well for another.

4. Perform new movements slowly. Walk the athlete through the required body positions, pausing wherever appropriate. Use an oral count for rhythm. Increase the speed of performance slowly. Always be prepared to repeat a step in this progression if speed reintroduces errors.

5. Attempt the complete skill at reduced speed. When you have decided that the athlete has adequately learned the actions you want to correct, put them back into the phase they came from and check the athlete's performance. If you are satisfied, attach additional phases from the skill at either end of the one containing the correction. Check how the athlete integrates the new movements. If problems persist, reinforce earlier steps in this sequence.

6. Progressively increase speed and effort.

Be aware that the time you have available to work with the athlete will considerably influence the error correction. Can you plan a training program that stretches over several months or a year, or are you working within a 3 to 6 wk block before the competitive season starts? Being restricted by time affects your choice of errors and the methods you use to correct them. You may think that all you can do is correct some minor errors because major changes initially cause the athlete to perform poorly and this downtime can carry into the competitive season. Whenever an athlete has to think about what he is doing in one or more phases of a skill, the effect is a drop in performance. The correct action must become second nature—an unconscious action. Remember that correcting a minor error can affect performance considerably. Be satisfied with that! Save large changes in technique for the off-season.

Communicate Ideas for Correcting Errors

How you communicate the ideas for change to the athlete determines how much success you gain when you attempt to correct errors. Don't befuddle a young athlete with needless technical jargon. Translate your mechanical know-how into instructions that fit the age, intelligence, and physical ability of the athlete. Some athletes will be genuinely interested in the mechanical principles behind a movement. Statements such as "Push forward when you contact the ball—it'll help you apply more force" are excellent because they indicate in easy-to-understand language the mechanical reasons behind a body position.

But for other athletes, comments about angular momentum, momentum transfer, kinetic energy, and rotary inertia are meaningless. Like people in general, athletes are different and will respond to different cues and comments. Give simple, short, easy-to-understand instructions that are to the point (Camiré and Trudel 2014). No athlete wants to stand around while the sport staff talks endlessly about what could or should be done.

In addition, during whatever process of correcting errors you use, remain positive and praise good effort and correct performance. Your objective is to help the athlete persevere through those difficult periods when she thinks that she will never master the correct action. The progress that the athlete makes will depend on the amount of practice performed, the complexity of the required actions, and the time required for you and the athlete to mold the corrected action into the total skill.

AT A GLANCE

Sequence and Process for Correcting Errors

- When correcting an error provide tips that translate your sport mechanics knowledge into language that the athlete can relate to.

- To a high-performance athlete, minor errors can make the difference between a good performance and a world record. When working with a beginner, minor errors are not as important as fundamental major errors.

- Some factors in the performance can be modified or corrected, such as takeoff angle or body position, whereas others factors such as limb length cannot be modified. Therefore, the underlying rule when determining which errors to work on is to focus on the ones that can be modified.

Use Outside Sources to Aid in the Correction

Few sports use mirrors in the way that bodybuilding does. Besides providing bodybuilders with continuous aesthetic assessment, mirrors provide instant visual feedback on how they perform an exercise. In the highly dynamic actions of most sports, the nearest you can get to a mirrored image is an immediate playback on video. The video tells the athlete, "This is what you looked like." But neither the video nor a mirror can say, "This is what the movements should feel like."

The athlete is the only one who actually feels the movement. As a sport professional observing and assessing the skill, you cannot feel what the athlete feels, although you can describe to the athlete what it should feel like. To provide this kind of information, you need considerable experience, first as an athlete and then as a sport professional.

Sometimes you'll be surprised at the response when you ask, "What did you feel at that moment in the skill?" Many novices have no idea what happened, and they have no sense of where their limbs are as they attempt a skill. High-performance athletes differ considerably. Most of them have a well-developed kinesthetic sense and are aware of what their bodies are doing during a performance. Teach young athletes to develop this sensory awareness. This resource becomes invaluable in helping them master sport skills.

In addition, you must gather information from many sources to further your knowledge in this profession. In the area of analysis and skill correction, you should be ready to do the following:

- *Research*. Read texts on your sport that offer successful teaching methods and techniques for correcting errors. The best texts include excellent illustrations of lead-ups and teaching progressions. A quality text contains lists of common errors that occur in the skills of your sport and explanations of how to correct them. Look also for safety recommendations, not only for individuals but also for group activities. This point is particularly important when you are teaching skills that have a high element of risk.

- *Attend coaching seminars and sport science workshops*. Here you can listen to presentations on coaching from experienced coaches and sport science experts who do research in your area. You can discuss problems in your sport with coaches who are aware of the latest coaching techniques. Join your local and national coaching associations or sport science associations so that you regularly receive the latest newsletters.

- *Access current technology*. Video and computer technology now gives you the opportunity to analyze performances in superslow motion and to place high-performance athletes on the same screen as novices. This approach is of tremendous

APPLICATION TO SPORT

Is the Methodology for Correcting Errors Transferable Across Sports?

Debate is occurring in the sport science community about the concept of generating lift when an object, like a rowing blade, moves through a fluid. Numerous studies have modeled various sporting implements (such as blades), as well as human hands (for swimming), in an attempt to determine the type and amount of forces generated during movement through a fluid. Using the well-established theories of airflow over an airplane wing that enables the plane to fly, questions arise about whether this principle applies in a sport like rowing.

Photodisc/Getty Images

For example, when a blade moves through the water, is this flow turbulent or laminar (because the type of flow will influence the lift and drag resultant forces)? Perhaps a better approach in applied sport mechanics is to accept that the principle of lift must exist (because this mechanical process is proven) and to focus the question on how much the lift forces contribute to the propulsion of the rower. As a professional working in the sport, you can manipulate the angle of attack of the implement (or hand) and see what the outcomes of propulsive forces are. Remember that only small adjustments in the angle of attack will be required and that these changes can alter the flow around and over the object.

assistance when you want to look at differences in technique. Computer advances in virtual reality now offer you and the athlete a "virtually real" method of experiencing the movement patterns of a skill while remaining static. Previously, downhill skiers and luge competitors would close their eyes and imagine steering through the curves and straightaways. With virtual reality the athlete can put on a specially designed helmet that feeds a three-dimensional image containing an accurate representation of the bends, twists, and straightaways that exist on the course. When applicable, a computer will have worked out the best course to follow relative to temperature and other weather conditions.

• *Observe other sports.* Read beyond this book and improve your knowledge of sport mechanics as it applies to your sport. Take an interest in other sports as well. Doing so will make you more of an expert in your own!

SUMMARY

- Five steps are used for effective observation, analysis, and correction of errors in sport skills: (1) Observe the complete skill, (2) analyze each phase and its key elements, (3) use your knowledge of sport mechanics in making an analysis, (4) select errors to be corrected, and (5) decide on appropriate coaching techniques for the correction of errors.

- Ensure safety when observing skills and observe skills from several positions. Avoid settings in which you and your athlete are distracted.

- Use technology such as video recordings to assist in your analysis and to provide feedback to the athlete.

- Use all your sensors (eyes, ears, and so on) to look for clues in an activity.

- When spotting, focus only on the safety of the athlete; do not concentrate on skill analysis.

- After you have a good overall impression of an athletic performance, analyze each successive phase of the skill together with its key elements.
- Use your knowledge of sport mechanics to analyze performance. Ask yourself mechanical questions (e.g., Does the athlete have maximum stability when applying or receiving force?).
- Generate a flowchart of what is required in the performance.
- Divide the performance errors you see into major and minor categories. Major errors seriously detract from the optimal performance of the skill, whereas minor errors have minimal effect on performance. Follow the sequence outlined in this chapter for correcting errors.
- Use the best safety techniques when correcting errors in skills that involve a high level of risk.
- Maintain a positive attitude during the correction process. Avoid using excessive technical jargon during coaching sessions.
- Teach athletes to develop sensory awareness to assist them in error correction.
- Be aware of the time you have available for correcting errors. Do not attempt massive changes in technique if time is limited.
- Attend coaching seminars and read texts on your sport that offer top-quality teaching methods and techniques for the correction of errors. Be alert to any advances in computer and video technology that can assist you in coaching. Expand your knowledge of sport mechanics, not only in your sport but also in other sports.

KEY TERMS

feedback

key elements

observation

phases

sequence

REFERENCES

Buszard, T., M. Reid, R. Masters, and D. Farrow. 2016. "Scaling the Equipment and Play Area in Children's Sport to Improve Motor Skill Acquisition: A Systematic Review." *Sports Medicine* 46 (6): 829-43.

Camiré, M., and P. Trudel. 2014. "Helping Youth Sport Coaches Integrate Psychological Skills in Their Coaching Practice." *Qualitative Research in Sport, Exercise and Health* 6 (4): 617-34.

Dingley, A. A., D.B. Pyne, J. Youngson, and B. Burkett. 2015. "Effectiveness of a Dry-Land Resistance Training Program on Strength, Power, and Swimming Performance in Paralympic Swimmers." *Journal of Strength and Conditioning Research* 29 (3): 619-26.

McKean, M.R., and B. Burkett. 2010. "The Relationship Between Joint Range of Motion, Muscular Strength, and Race Time for Sub-Elite Flat Water Kayakers." *Journal of Science and Medicine in Sport* 13 (5): 537-42.

Sinclair, J., P.J. Taylor, S. Atkins, and S.J. Hobbs. 2016. "Biomechanical Predictors of Ball Velocity During Punt Kicking in Elite Rugby League Kickers." *International Journal of Sports Science and Coaching* 11 (3): 356-64.

Zemková, E. 2014. "Sport-Specific Balance." *Sports Medicine* 44 (5): 579-90.

 Visit the web resource for review questions and practical activities for the chapter.

Selected Sport Skills

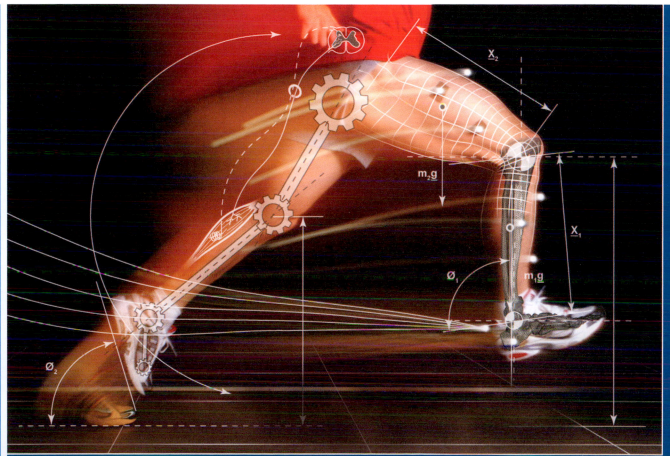

Courtesy of Biomechanigg Sport & Health Research Inc, www.biomechanigg.com.

When you finish reading this chapter, you should be able to explain

- how to apply the sport mechanic principles to the common types of sports;
- how to transfer the mechanical knowledge into technique;
- how to break the sport into key specific phases; and
- how to identify the key components that need to be addressed when performing these sports.

In this final chapter of the text we analyze a number of sport skills to show you how technique and the mechanics of sport are inseparable. You'll find that the following pages are a good review of the mechanical principles you read about earlier in this book. This chapter combines the fundamental principles of sport mechanics with core sports in a truly applied nature. These principles can be grouped into four themes:

1. Movement of the athlete (sprinting, jumping, and wheelchair sports)
2. Movement of implements (throwing, striking and batting, and swinging and rotating)
3. Movement of an opposition (weightlifting, combat throwing, and tackling in football)
4. Movement through a fluid (swimming, kicking a ball, paddling a watercraft)

The technique and sport mechanics of 12 sample sport skills are described on the following pages. Table 13.1 lists the sport skills you will find in this chapter and the rationale for including them.

Each of the sport skills is explained in detail, as you will see later in the chapter. In a corresponding table under the heading "Technique," you'll find the most important technical characteristics of the skill—the movement patterns that should occur

TABLE 13.1   Rationale for Selection of Skills in 12 Sports

Sport skill	Rationale
Movement of the athlete	
Sprinting	Sprinting is the most dynamic and vigorous running technique. Walking and middle- and long-distance running obey the same mechanical laws as sprinting but use less vigorous actions.
Jumping	The mechanical principles that control how a jumper gets up in the air in the high jump also apply to other jumping skills (e.g., a volleyball spike, a basketball layup, and a receiver's leaping catch in football). Once in flight, any athlete is subject to the same mechanical laws as the high jumper is.
Wheelchair sports	For athletes with a disability, the wheelchair is a common assistive device that enables them to get a move on. The sport of wheelchair tennis, for example, requires the skill of generating high accelerations and the unique incorporation of the athlete–equipment interface.
Movement of an implement	
Throwing	Maximizing the distance of a throw requires the athlete to generate angular rotation as fast as possible throughout the body, culminating with the thrower's hand moving as fast as possible. Throwing skills rely on an effective kinetic link.
Striking and batting	To strike or bat an object effectively, the athlete must rotate the implement as quickly as possible. This action requires maximization of the mechanical principle of angular velocity, which is generated by effective implementation of levers. For striking and batting skills, athletes attempt to simulate a whiplash action with their limbs to generate maximum velocity of the swinging implement.
Swinging and rotating	Most swinging and rotation events occur in the sport of gymnastics. For example, a gymnast's back giant on the high bar is a rotary swinging motion performed around a high bar, which acts as an axis external to the gymnast's body. In contrast, the axis for a front somersault passes through the athlete's body from one hip to the other. The mechanics of these two skills include both similarities and differences. To generate and control swinging and rotation, the athlete must manipulate the length of the radius of gyration to influence the angular rotation.

Sport skill	Rationale
Movement of an opposition	
Weightlifting and powerlifting	In weightlifting, the clean is a lifting–pulling action and the jerk is a push. Carrying and supporting actions occur when the weightlifter pauses with the bar at the chest and again when the bar is at arm's length above the head. The mechanical principles involved in a weightlifter's clean and jerk apply to all lifting, carrying, and spotting techniques. The laws controlling stability also play an important role in the clean and jerk.
Combat throwing	Combat sports require athletes to maintain their stability and "throw" the opposition. The fundamental mechanical battle occurs when athletes try to maintain their own stability and at the same time destroy the stability of their opponents. These mechanical principles of stability, governing the front giant, the front somersault, and the clean and jerk, appear again in the judo hip throw.
Tackle in football	An effective football tackle requires the athlete to generate a solid and stable base of support. During tackling, the line of gravity and base of support are essential, just as they are in weightlifting and judo.
Movement through a fluid	
Swimming	The freestyle stroke is the most popular and fastest of all major swimming strokes. The mechanical principles of drag and lift that govern the technique used in freestyle apply to other swimming strokes, as well as to the sculling actions used in water polo and synchronized swimming.
Kicking a ball	To propel the football through the air, the athlete must generate maximum velocity when the foot makes contact with the ball. After the ball is kicked, it is subject to the mechanical principles of projectile motion, the same as for any object, such as a discus or javelin.
Paddling a watercraft	The fluid of water creates unique challenges of large drag forces for the athlete, particularly for those in higher-velocity sports that involve equipment such as a rowing skull or kayak.

when athletes sprint, swim, throw, hit, jump, lift, and so on. In those same tables, under the heading "Mechanics," you'll find a description of mechanical principles at work during the performance of the skill's technique. Here you'll find the mechanical reasons that a skill, and each of its phases, is performed the way it is.

This analysis does not cover every sport skill that exists, nor will it satisfy the needs of every professional working in every sport. But we've tried to cover a wide range of sports that represent the core sport skills. If this review highlights for you how technique and sport mechanics are intrinsically tied together and why you cannot teach technique without knowing fundamental mechanics, then it's done its job!

As you read the following pages, look for mechanical principles that constantly reappear from one skill to the next. Watch for the turning effect of torque, the battle against inertia as athletes accelerate, and athletes' use of impulse. Whatever skill you consider, these mechanical principles (and many others) are always present. Get to know them, recognize them, and understand their effect on athletic performance. This process will help you immensely in your profession in the sport industry.

Sprinting

The highlights of sprinting, as shown in figure 13.1a through 13.1e, are as follows:

• The key phases in sprinting are the flight phase and the stance phase. As a continuous cyclic action these phases constantly repeat. The proportion of time within each of these stance and swing phases will change with sprinting velocity; for example, a faster velocity will increase the flight time.

• The time that an athlete takes to run a set distance depends on the athlete's stride length and stride frequency. The length of the athlete's legs and the forward thrust that occurs with each stride determine stride length. Forward thrust is produced by the earth's reaction force responding to the athlete's backward thrust against the earth's surface. Stride frequency is the cadence that the athlete uses (i.e., the number of strides that occur each second).

• A runner's technique changes as the athlete runs faster. Sprinters spend more time in the air than distance runners do. In addition, they flex and swing their arms more vigorously. Sprinters

a   *b*   *c*   *d*   *e*

FIGURE 13.1   Sagittal plane view of sprinting technique for one step.

also have a higher knee lift, a greater leg thrust, and a higher-flexed leg. Distance runners use less arm action but tend to swing their shoulders more than sprinters do. The longer the distance is, the greater the reliance is on cardiovascular endurance and pacing. All high-performance runners hold their torsos close to perpendicular, with a slight forward lean.

• Tension is detrimental to all runners because tension saps energy and restricts muscle action and limb movement. Sprinters try to run explosively while still relaxing their faces, necks, shoulders, and hands. Distance runners use the same relaxation techniques.

• Training optimizes the forward thrust that occurs at each stride. Training also brings more muscle fibers into action and teaches the athlete to relax opposing muscle groups. Leg drive is improved by related power training and flexibility through increased range of movement.

Sprinting Characteristics and Mechanical Principles

Technique	Mechanics
Good sprinting technique demands an optimal blend of stride length and stride frequency. A predominance of fast-twitch muscle fibers is essential for top-level sprinting.	A combination of optimal leg power, stride length, and stride frequency produces the best sprinting times. Power, good reactions, and excellent flexibility are all essential. Stride length depends on hip flexibility, leg length, muscle power, and range of movement. Overemphasis on stride frequency (i.e., stride rate or cadence) or stride length produces inefficient sprinting.
Sprinting requires excellent leg, hip, and shoulder flexibility. Flexibility in the hip and pelvic area is particularly important.	An ability to rotate the hips around the long axis of the body helps produce optimal stride frequency and stride length. Flexibility in the shoulder girdle promotes good arm swing.
A sprinter's arms are comfortably flexed at 90° and swing powerfully forward and backward. The hands are relaxed and swing hip high to the rear and shoulder high in front (see figure 13.1*a*-13.1*e*).	Forward and backward arm swing counterbalances the twisting motion produced by each leg thrust on either side of the sprinter's long axis. Flexing the arms at the elbows reduces their rotary inertia and makes their pendulum movement easier for the muscles involved. The forward swing of each arm transfers momentum within the athlete's body. This action adds to the athlete's leg thrust and helps drive the athlete forward. Forward and backward arm swing parallel to the direction of sprint (rather than across the body) helps hold the torso and the shoulder girdle steady. This action aids balance and relaxation and ensures that the athlete runs a straight line toward the finish.

Technique	Mechanics
The driving leg extends to near full extension (see figure 13.1b). When the driving foot leaves the ground, this leg flexes at the knee, bringing the heel up to buttock level (see figure 13.1e).	A powerful leg extension through the hip, knee, and ankle joints provides the athlete with optimal thrust in the direction of the sprint. Thrust backward and downward around 50° to 55° produces an equal and opposite reaction from the earth, which drives the athlete in a predominantly horizontal direction along the track. Flexion of the legs (like that of the arms) reduces their rotary inertia and makes their recovery and forward movement easier for the muscles involved.
After thrusting backward and downward, the driving leg flexes at the knee and is brought directly forward and upward so that the thigh swings to just below horizontal (see figure 13.1c).	The swing and upward thrust of the driving leg as it is brought forward is counterbalanced by the action of the opposing arm. Forward thrust of both arm and leg increases the momentum transfer. This action helps produce a greater thrust back at the earth with the driving leg, and in response, from the earth propelling the sprinter's body along the track.
When the sprinter's driving leg is recovered and becomes the supporting leg, it flexes slightly on landing. The supporting foot lands in front of the sprinter's center of gravity. The first contact of the foot with the ground is typically on the outside edge of the foot. Depending on running velocity, the heel is lowered but does not contact the ground.	A slight flexion of the supporting leg extends the time of impact during which force is applied to the sprinter's body and therefore cushions the landing. Flexion lengthens the leg muscles, which are then ready to extend the driving leg backward and downward against the earth. The landing of the supporting foot in front of the sprinter's center of gravity creates deceleration, but the gains in stride length counter this issue (see figure 13.1).
A sprinter's forward body lean is extreme during a sprint start. At top speed, the torso is close to perpendicular and the shoulder girdle is held square to the direction of run (see figure 13.1a-13.1e).	During a sprint start, forward body lean and shorter high-frequency strides overcome the resting inertia of the sprinter's body mass and help the sprinter gain momentum. At full speed, a perpendicular torso coupled with vigorous forward and backward arm action counterbalances the movement of the legs.
A sprinter's body rises and falls very little during running at full speed (see figure 13.1a-13.1e).	A high-performance sprinter's center of gravity follows a low wave-like pattern as it travels forward. Slightly more time is spent in the air than in a support position. Too much time in the air is time wasted and indicates that too much thrust is directed in a vertical direction.
The sprinter's head is held in a natural alignment with the torso. Vision is horizontal and directly ahead (see figure 13.1a-13.1e).	Proper position of the head and vision assists in maintaining the stability of the sprinter's torso. Tilting the head back can increase tension and restrict stride frequency and stride length.
Good sprinting combines power with relaxation. Face, neck, shoulders, and in particular the hands are relaxed.	Tension in the body can reduce the velocity of muscle contraction and therefore reduces sprinting velocity. Good sprinting requires a rapid change from muscle contraction to relaxation. A technically superior sprinter is mechanically efficient and relaxed. Unnecessary tension is avoided, and in this way the athlete uses energy efficiently.
An athlete's sprinting speed is influenced by environmental conditions.	The greatest expenditure of energy in sprinting occurs when the athlete is thrusting back at the earth. Energy is also required for the knee lift and support phase of sprinting. The faster the athlete runs, the more energy the athlete must spend fighting air resistance. Headwinds add to this resistance. The condition of the track can influence sprinting speed. Lightweight, high-quality spikes increase friction with the ground and therefore enhance traction. This provides a safer landing, and the ability to drive back on a firm rubberized track helps thrust the athlete forward better than running on soft ground. Friction-reducing running tights and slick bodysuits help minimize drag from the air.

Jumping

The following are the highlights of jumping, as in a high jump, shown in figure 13.2a through 13.2e:

• Jumping requires a combination of running velocity, muscle power, and precise coordination of body limb segments to jump at maximum height at a specific location. The run-in generates horizontal and vertical velocity before the flight phase. To get up in the air, jumpers exert a force against the earth's surface well in excess of their own body weight. The earth's reaction force in combination with the athlete's synchronized muscle contraction then drives the athlete upward. The more forceful the athlete's thrust against the earth is, the greater the earth's response is.

• The manipulation of the athlete's center of gravity is critical for an effective jump. Immediately before takeoff, a jumper's center of gravity is lowered, the body is tilted backward, and the athlete's arms and free leg are positioned to the rear of the body. Lowering the body prestretches the large and powerful muscles of the jumping leg, preparing

them for the leg's explosive thrust downward at the earth. Leaning back combines with lowering the body so that the athlete can spend more time over the jumping foot applying force to the earth. Swinging the arms forward and upward adds to the downward thrust of the athlete's jumping leg against the earth.

• The path that a jumper's center of gravity follows during flight is initially determined by the velocity at which the athlete is propelled upward at takeoff and the takeoff angle used. This flight path can be maximized by the effective placement of the upper and lower limbs.

• When in flight, movement of one part of a jumper's body causes other parts to move in the opposing direction (action and reaction). In the high jump, this characteristic helps in bar clearance. In the long jump, rotary actions of the arms and legs are used in flight to counteract the unwanted forward rotation that inevitably occurs when the athlete takes off. In a volleyball spike, drawing the arm back and arching the body in a counter-clockwise direction causes the legs to move in a clockwise direction.

FIGURE 13.2   Sagittal plane view of high-jump technique.

Jumping Characteristics and Mechanical Principles

Technique	Mechanics
The approaches used by high-performance athletes in the high jump range from 10 to 13 strides and commonly extend over a curved approach of 18 to 33 m (60 to 110 ft). Jumpers approach the bar at high velocity and accelerate to even greater velocity during the last three strides before takeoff.	A high-jump approach should be long enough that sufficient velocity is reached to carry the athlete through the actions performed at takeoff.

Technique	Mechanics
The greater the velocity of the approach is, the greater the potential is for helping the athlete jump high. Too little velocity in the approach can be detrimental to the takeoff, but too much velocity can also be detrimental.	The speed of the approach must be fast enough to carry the jumper through all the body positions required in the takeoff. Time must also be sufficient to generate the optimal vertical, horizontal, and rotary thrust to get up and over the bar. On the other hand, if the approach speed is slow, the athlete will move far too slowly through the takeoff and will have a difficult time getting from a back-leaning position at the start of the takeoff to an upward-rotating thrust that occurs just before leaving the ground.
Most high jumpers use a curved approach that has a large-radius curve (or is almost straight) during the first six or seven strides, followed by a small-radius curve during the final three to five strides. Jumpers lean into the curve of the approach. They also lean backward as they plant the takeoff foot (see figure 13.2a).	The athlete's inward lean during the approach produces a centripetal force. A tighter curve and greater approach velocity require more inward lean. Inward and backward lean as the jumping foot is planted lengthens the time that the athlete spends thrusting down at the earth with the jumping leg. Inward and backward lean before takeoff counterbalances the outward pull of inertia. A body position with lean away from the bar prevents the athlete from committing the error of leaning into (i.e., toward) the bar.
During the last two or three strides of the approach, the athlete's center of gravity is lowered and the arms and free leg are positioned to the rear of the body (see figure 13.2a). The penultimate (i.e., next to last) stride and the final stride are longer than previous ones. The athlete steps forward onto the jumping foot with the hips positioned well to the rear of the jumping foot.	Lowering the center of gravity and stepping well forward onto the jumping foot allow the athlete to apply force against the earth over a large arc (i.e., a long period) and in reaction to have the earth spend more time driving the athlete upward. Lowering the center of gravity prestretches the jumping muscles in preparation for their powerful extension of the jumping leg.
The arms and free leg are positioned to the rear of the body as the athlete steps forward onto the jumping leg (see figure 13.2a). The forward and upward swing of the arms and free leg is then coordinated with the extension of the jumping leg (see figure 13.2b).	The upward swing of the arms and the free leg toward the bar is a form of momentum transfer. The momentum of this upward swing combines with the downward thrust of the takeoff leg. All unify to produce a greater reactive response from the earth, which drives the athlete upward.
The free leg and arms flex and accelerate during their upward swing. These actions occur while the takeoff leg is extending vigorously in contact with the earth (see figure 13.2b).	Flexing the arms and legs brings their mass closer to their respective axes and reduces their rotary inertia. This action makes it easier for the athlete's muscles to move them upward at high speed. For greatest effect, the free leg and arms move at maximum velocity at that instant when the athlete is last in contact with the earth. The muscles in the jumping leg must have enough power to extend the jumping leg explosively.
The takeoff for elite high jumpers is from a point 0.9 to 1.2 m (3 to 4 ft) directly out from the near high-jump standard. The takeoff foot is placed at an angle of 15° to 20° to the crossbar. The free leg is first thrust upward and then rotated away from the bar and back toward the direction of approach. The shoulders rotate parallel to the bar. Vision is usually along the bar in the direction of the far standard.	Rotation of the jumper's body in preparation for a back-lying position over the bar is initiated while the takeoff foot is still in contact with the ground. The athlete is then able to rotate around the body's long axis by pushing against the earth's surface. In the air, the same movement causes a twist-like equal and opposite reaction to occur. Elite jumpers try not to compromise vertical thrust by overemphasizing rotation during the takeoff.
At takeoff, the high velocity of the approach coupled with the plant of the takeoff foot causes the athlete to move forward over the takeoff foot.	Moving up and over the takeoff foot initiates rotation, which continues in flight. Once in flight, the athlete can rotate more quickly (i.e., increase angular velocity) by pulling the body inward or slow down rotation by stretching out and extending the body.

> continued

Jumping Characteristics and Mechanical Principles   *> continued*

Technique	Mechanics
After takeoff the athlete's upper body is flexed backward over the bar. The free (i.e., leading) leg, which was swung up at the bar, is lowered from its elevated position (see figure 13.2c).	Lowering the leading leg, combined with the act of flexing the upper body backward, produces an equal and opposite reaction that elevates the athlete's hips. This action clears the hips over the bar. The athlete's vision, directed along the length of the bar, helps time this critical maneuver.
When the athlete's seat has crossed the bar, the head and shoulders are lifted upward. The athlete's torso and the legs (which are now flexed at the knee) are pulled toward each other by contraction of the abdominal and quadriceps muscles (see figure 13.2d).	The elevation of the head and shoulders causes two equal and opposite reactions to occur: (1) The flexed legs move upward toward the torso, and (2) the athlete's seat (which has crossed the bar) drops downward. Flexion of the legs reduces their rotary inertia, making their movement easier so that they are quickly drawn over the bar.
After the athlete's thighs have cleared the bar, the legs are extended at the knees so that the heels avoid clipping the bar (see figure 13.2e).	Even though the legs extend, the fact that the athlete's upper and lower body have moved toward each other causes the athlete's body to increase its rate of spin (angular velocity) around the athlete's transverse (hip to hip) axis. This action, combined with some rotation around the long axis, continues as the jumper drops toward the landing pad.
The athlete relaxes and drops onto the shoulders on the high-jump landing pad.	The continued rotation of the athlete's body in flight causes the athlete to drop onto the shoulders on the landing pad. The focus of the eyes on the bar during the clearance of the feet pulls the chin into the chest and prevents a headfirst landing. The height of the high-jump landing pad relative to the bar is designed to stop overrotation and prevent the athlete from landing on the back of the neck. The foam rubber of the landing pad extends the period and area of impact and gradually reduces the forces applied to the athlete during the landing.

Wheelchair Sports

The highlights of wheelchair sports, such as wheelchair tennis, as shown in figure 13.3, are the following:

• A wheelchair is a device used for mobility by an athlete for whom walking is difficult or impossible because of illness or impairment. A wheelchair typically consists of a seat supported on two large wheels on an axle attached toward the back of the seat and two small wheels near the feet, although small additional features are often included to prevent toppling or to assist in mounting curbs. The athlete moves by using the hands to push on circular bars on the outside of the large wheels that have a diameter slightly less than that of the wheels, or by actuating motors, usually with a joystick.

• Manual wheelchairs with push rims enable athletes to propel themselves in the chair. The push rims also allow change of direction and in some cases the ability to lift the front wheels of the chair off the ground. To avoid damage to the hands, because they may get caught within the spokes of the rim, chair users may wear fingerless gloves such as weightlifting gloves; these protect the hands from both dirt and injury. Chairs without push rims are usually pushed by a person using handles behind the chair and are typically not used for sport because they are not self-propelled.

• For activities or sports that are generally linear (such as track racing), high-performance athletes with an impairment, Paralympians, may use purpose-built streamlined three-wheeled chairs. These devices have evolved to meet the specific requirements of the sport.

• Rolling resistance is the first thing to overcome to propel a wheelchair. The rolling resistance depends on the surface that the wheelchair is driving on, mass distribution on the wheels, wheel radius, total mass, and specific tire characteristics. The most important external aspect in all this is the surface on which the wheelchair is moving. Indoors, the floors may be hard and smooth. Outdoors (such as uneven ground, gravel, or broken surfaces), the characteristics of the surface increase resistance.

• The athlete's technique of pushing the wheelchair can be assessed with several devices, such as a velocometer (Tolfrey et al. 2012). These devices provide objective feedback on how effective the athlete is pushing the wheelchair and can quantify any modifications to the athlete's pushing technique.

AT A GLANCE

Movement of the Athlete

- Typically, the movement of an athlete is the product of the athlete's stride or push length and stride or push frequency.

- The manipulation and orientation of the athlete's center of gravity plays a significant role in the effectiveness of the athletic performance.

- In wheelchair sports the sport-specific design of the wheelchair can influence the effectiveness of the performance, along with the surface that the wheelchair is moving on.

FIGURE 13.3 Frontal plane view of tennis-specific wheelchair.

Moto Yoshimura/Getty Images

Wheelchair Sports Characteristics and Mechanical Principles

Technique	Mechanics
Wheelchair tennis is one of the official Paralympic sports and is played at Grand Slams and Paralympic Games. Each of the three categories—men, ladies, and quads—has singles and doubles tournaments. Quads was originally established for those with quadriplegia, but it includes spine, hip, knee, and ankle impairments and other lower-limb impairments. Quads players can hold rackets taped to the hand and use electric-powered wheelchairs.	Wheelchair tennis is one of the forms of tennis adapted for those who have impairments in their lower bodies. The sizes of courts, balls, and rackets are the same, but wheelchair tennis differs in two major ways from pedestrian tennis: The athletes use specially designed wheelchairs, and the ball may bounce up to twice. The second bounce may occur outside the field. This rule is one of the reasons that wheelchair tennis has become popular—people in a chair can easily play against able-bodied players.
The athlete needs to be comfortably and safely seated in the chair.	The position of the seat influences the accessibility of the hand rim and therefore the efficiency of power transmission from hand to hand rim and the mechanical efficiency.
More advanced players may use wheelchairs specifically designed for tennis; these have a single small wheel at the front and back of the chair to facilitate stability when turning left or right. To increase stability, athletes may also use wide Velcro straps to hold themselves in the chair.	The most important aspects of the seat of a wheelchair are the horizontal and vertical positions of the user, because they greatly influence the energy needed to propel the wheelchair. In general, positioning the center of mass right above the rear-wheel axle (horizontal position) is best. In the vertical direction, users should be positioned in such a way that they can just touch the rear-wheel axle with their fingertips.

> continued

Wheelchair Sports Characteristics and Mechanical Principles   > *continued*

Technique	Mechanics
To propel the wheelchair, the user needs to grip the push rim or the wheel or both. To go forward, the user reaches back as far as possible and grips the wheels. The user pulls and then pushes the wheels forward by keeping hold of the handles until the end of the stroke.	The way of grabbing the hand rim when propelling influences the mechanical efficiency greatly. Also, the friction coefficient is of great influence. It should be as low as possible to avoid braking the wheels during propulsion, but it should be high enough to make it possible to transmit a certain amount of power from hand to hand rim.
Turning and steering the wheelchair is a combination of gripping one side of the chair and pushing the other. To stop, the user grips the rims and uses friction to slow them down. Users who wear gloves can press the palms of the hands against the wheel rims to create friction. This technique keeps the fingers safer.	To turn right, users hold the right wheel still and push the left wheel forward. To turn left, they hold the left wheel still and push the right wheel forward. If they need to spin on the spot or in a tight corner, they push one wheel forward and the other one backward simultaneously. The key to gripping is the friction between the hands or glove) and the rim of the chair. Users can increase the friction by changing the texture of the gloves.
The diameter of the hand rim can be changed depending on the anthropometric profile and the available strength of the athlete. Users must be able to reach the ground while still sitting in the wheelchair; otherwise they will not be able to pick up the tennis ball.	A large diameter of the hand rim results in a relative high mechanical efficiency and effective force. When propelling a wheelchair over a long distance, using a hand rim with a smaller diameter is energetically favorable. For wheelchair sports like tennis in which the player needs to hold the tennis racket in one hand and push the chair at the same time, a larger wheelchair rim is advantageous.
A camber angle of the wheel has a positive influence on the stability sideways, the power transmission from hand to hand rim, and the maneuverability of the chair.	Because of the wheel camber angle, there is more risk for toe-in or toe-out and more pressure on the rear-wheel axle. In addition, the complete wheelchair becomes wider. In general the camber angle for a wheelchair for activities of daily living is 2° to 4°; for a racing wheelchair it is between 4° and 12°.
Wheelchair users should avoid going over bumps where possible, because hitting a bump (even one as small as 1 cm [.5 in.]) at speed can catapult them out of the chair and across the floor. They should go slowly over bumps and if possible go backward to improve control.	Because of the unstable movement of the wheels, going over bumps requires a controlled center of gravity before, during, and after the bump. Users should lean slightly forward during takeoff to move the center of gravity closer to the bump. In midair, they need to correct by steadying themselves a little bit backward so that the back wheels hit slightly before the front wheels do. They need to be careful not to fall backward.

Throwing

The highlights of throwing, as in a javelin throw, shown in figure 13.4, are the following:

• High-velocity throwing skills require the athlete to coordinate the movement of her anatomical segments before transferring to the implement, similar to the cracking of a whip. The high-speed movement of a javelin throw requires the athlete to generate phenomenal velocity at the point of release of the implement. This flail-like action is called a kinetic link or kinetic chain action because the slower motions produced in the athlete's longer, larger, and heavier limbs are made faster and quicker as they pass their motion on to lighter, less massive body parts.

• In a high-velocity throw, such as a javelin throw, a rapid acceleration of the athlete's body segments occurs, beginning with those in contact with the earth. A whiplash, or flail-like, sequence progresses along the kinetic chain, upward from the legs to the hips and then from the hips to the chest. The action culminates in the tremendous velocity of the striking or throwing arm.

• To achieve the greatest possible velocity, antagonist muscle groups must be completely relaxed and the agonist muscles must contract in sequence, helping to make each body segment move faster than the previous one.

• Each movement pattern in throwing skills contains preparatory actions, commonly called approach, setup, and windup or backswing. These actions help place the body and implement (such as a javelin) in the optimal position for the application of force. A backswing (such as in a golf drive) provides additional distance over which the club can be accelerated, thereby generating a faster club-head velocity at impact. The backswing prestretches the golfer's muscles in preparation for their explosive recoil. Relaxation and flexibility help produce an optimal windup.

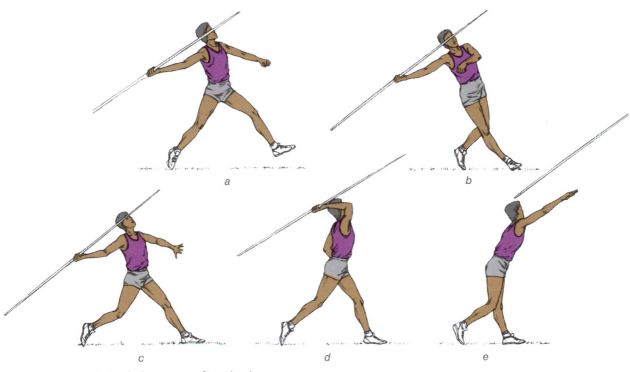

FIGURE 13.4   Sagittal plane view of javelin throw.

Throwing Characteristics and Mechanical Principles

Technique	Mechanics
The throw begins with an approach in the direction of throw because this leads into a wide, powerful throwing stance. The approach is controlled, begins slowly, and increases in velocity. The athlete accelerates during the first two-thirds of the approach. This action leads into the final one-third of the approach, which includes drawing back the javelin and stepping into the final throwing stance.	The approach builds momentum and generates enough velocity to carry the athlete through the throwing stance and into the follow-through. Too much velocity in the early part of the approach may cause the athlete to slow down during the throwing actions or may not allow enough time in the throwing stance to apply an optimal amount of force to the javelin. Efficient use of an approach can increase the distance thrown by 27 to 30 m (90 to 100 ft) compared with a standing throw.
Before entrance into the throwing stance, the athlete's shoulder girdle is rotated away from the direction of throw and the throwing arm is taken back to arm's length (see figure 13.4a). By use of one or more crossover steps, the lower body moves forward under the torso so that the athlete's body is angled backward away from the direction of throw (see figure 13.4a-13.4b).	Rotating the shoulder girdle and extending the throwing arm prepare the athlete for the application of force to the javelin over the largest possible distance and time. The backward body lean makes this distance and time even greater.

> continued

Throwing Characteristics and Mechanical Principles   *> continued*

Technique	Mechanics
The athlete steps into the throwing stance with the leg on the side of the body opposite the throwing arm. This step is usually larger than those preceding it (see figure 13.4c).	Stepping forward with the opposing foot (e.g., the left foot if the javelin is held in the right arm) sets up a large base of support for the application of force to the javelin. This stance allows the athlete's hips and shoulders to be rotated back toward the approach and away from the direction of throw.
The athlete's body is tilted backward, and the center of gravity is lowered over a partially flexed rear leg. The rear leg is flexed at the knee and angled outward 45° from the direction of throw.	Flexing the rear leg stretches the leg muscles in preparation for their explosive rotary thrust toward the direction of throw. This rotary motion is the first stage in the whiplash action that starts from the athlete's base (i.e., where the athlete's feet are in contact with the ground) and progresses up to the throwing arm.
The rear leg is vigorously rotated toward the direction of throw. This action thrusts the hips in the same direction (see figure 13.4c-13.4d). The muscles joining the hips to the torso stretch and contract explosively.	More massive, slower-moving parts of the body shift forward into the throw while lighter body segments (e.g., the throwing arm) complete their backward extension. This motion stretches the muscles in the abdomen, chest, and shoulders, getting them ready for their explosive contraction during later phases of the throw.
The athlete's torso rotates and pulls the shoulders and the throwing arm toward the direction of throw. Opposing muscle groups are relaxed. As the shoulders are pulled forward, the muscles of the shoulders stretch and then contract vigorously. A relaxed throwing arm follows the shoulder with a flail-like action. The athlete's body tilts sideways, away from the throwing arm. The free arm rotates backward to help pull the chest and throwing arm around and into the throw (see figure 13.4c-13.4d).	Each of the athlete's body segments, from the legs through to the shoulders and throwing arm, sequentially accelerates. This sequence sets up a flail-like whip-cracking action that progressively builds and ends in the tremendous velocity of the throwing arm. A sideways inclination of the athlete's torso allows greater height of release. The free (i.e., nonthrowing) arm is pulled backward to help rotate the athlete's torso around the long axis of the body. Rotation of the torso makes a contribution in pulling the throwing arm at high velocity into the throw.
As the athlete's throwing arm is pulled forward, the upper arm and elbow lead; the throwing hand and javelin trail well behind. Flexion occurs at the elbow of the throwing arm (see figure 13.4d).	Flexing the throwing arm at the elbow serves two purposes: (1) It helps the movement become even more whip-like, and (2) the elbow acts like the axle of a wheel with the throwing hand rotating around at its rim. This wheel–axle arrangement increases the velocity of the throwing hand and the javelin.
The athlete thrusts forward toward the direction of throw. The torso moves forward beyond the supporting leg, which has been straightened (see figure 13.4e).	Forcefully driving the body as far as possible toward the direction of throw extends the application of force to the javelin over the longest possible distance and time.
The athlete uses a follow-through to complete the throw (see figure 13.4e).	The follow-through applies force to the javelin for as long as possible. After the javelin has left the athlete's hand, the follow-through allows for safe dissipation of momentum from the athlete's body.
The angle of release of the javelin varies according to the throwing ability of the athlete, the type of javelin used, and environmental conditions at the time of throwing.	A javelin is dramatically affected by lift and drag forces. The trajectory angle and the angle of attack at release relative to environmental conditions, such as head- and tailwinds, determine the path and the distance of the javelin's flight. A moderate headwind of 8 to 16 km/h (5 to 10 mph) provides excellent lift to the javelin because it pushes up on the undersurface of the javelin. A tailwind is detrimental because it pushes down on the upper surface of the javelin.

Striking and Batting

The following are the highlights of striking and batting, such as swinging the bat in baseball, as shown in figure 13.5:

• Batting or striking skills require a well-co-ordinated, controlled but fast swing, because the implement (such as a baseball bat) must make contact with the smaller and fast-traveling ball. The high-speed movement of a 150 km/h (95 mph) fastball in baseball requires the athlete to generate phenomenal velocity with fine motor control to place the bat in the right position. The velocity at the bat is generated through the kinetic link, or kinetic chain action, because the slower motions produced in the athlete's longer, larger, and heavier limbs are made faster and quicker as they pass their motion on to lighter, less massive body parts. The process for a baseball swing is similar to that for movements in other sports, such as the tennis serve, the golf drive, and the badminton smash.

• Maintaining accuracy of a swing first requires the establishment of a stable base; the position of the feet and the associated line of gravity are critical to the overall success of the swing. A stable base allows a repeatable swing characteristic. Generating the high velocity of the implement requires rapid acceleration of the athlete's body segments, beginning with those in contact with the earth. A whiplash, or flail-like, sequence progresses upward from the legs to the

hips and then from hips to chest. It culminates in the tremendous velocity of the arm or arms (in the case of a baseball swing or a double-handed backhand in tennis).

• To achieve the greatest possible velocity, antagonist muscle groups must be completely relaxed and agonist muscles must contract in sequence, helping to make each body segment move faster than the previous one. This functional anatomy control also allows an accurate swing pattern.

• Each movement pattern in striking and batting skills contains some preparatory actions commonly called approach, setup, and windup or backswing. These actions help place the body and implement (such as a bat in baseball pitching) in the optimal position for the application of force. A backswing in a golf drive provides additional distance over which the club is accelerated. The backswing prestretches the golfer's muscles in preparation for their explosive recoil. Relaxation and flexibility help produce an optimal windup.

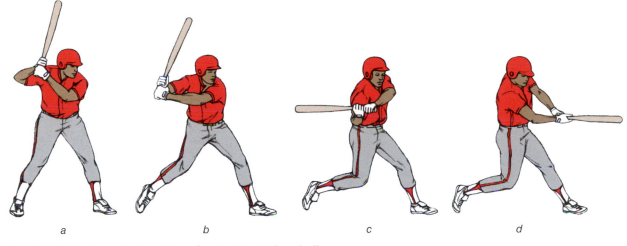

FIGURE 13.5 Frontal plane view when batting in baseball.

Striking and Batting Characteristics and Mechanical Principles

Technique	Mechanics
In baseball, a batter faces an approaching ball that is usually spinning and varies in velocity and direction. A baseball bat is cylindrical and therefore has a curved striking surface. Hitting the ball effectively is extremely difficult. Fastballs can have velocities of close to, and even over, 160 km/h (100 mph). Batters have less than a second to react to the pitch.	When a batter wants to strike a ball and send it in the opposing direction, the momentum of the bat must be greater than that of the ball. Arm extension, bat length, and the rate of swing determine the velocity of the barrel end of the bat. Timing and coordination are essential for hitting the ball through its center of gravity rather than hitting it off center (i.e., out of line with the ball's center of gravity) and producing a pop-up.

> *continued*

Technique	Mechanics
Batters take a stance that is slightly wider than shoulder-width. Their body weight is close to, or over, the rear foot, which is at right angles to the approaching pitch. Right-handed batters stand with the left side and left hip toward the pitcher. The head is turned so that the batter can concentrate on the pitcher's actions. The bat is frequently held in a ready position, with the barrel end pointing skyward (see figure 13.5a).	The initial shoulder-width stance with the body weight above the rear foot allows the batter to shift the center of gravity forward an optimal distance into the hitting stance. With the left side facing the pitcher, right-handed batters can rotate their hips and torsos more than 90° as part of the batting action. Moving the body toward the pitch (as well as rotating the hips and torso) increases the distance over which force is applied to the bat and subsequently to the ball.
As the ball approaches the plate, right-handed batters step approximately an additional half shoulder-width toward the pitcher. This movement puts them into a wide, powerful hitting stance. The batter's arms extend partially away from the body. Some backward rotation of the shoulders and the bat often occurs as the batter shifts forward toward the ball (see figure 13.5b).	Movement of the batter's body mass toward the pitch stretches the muscles that subsequently pull the bat around toward the ball. The trailing action of the shoulders, arms, and bat in relation to the movement of the batter's hips sets up a whiplash action characteristic of high-velocity throwing, kicking, and striking skills. Body rotation, coupled with arm extension, increases the angular velocity of the bat.
As the batter (right-handed) shifts forward from the right rear foot toward the left, the hips rotate around an axis formed by the left side of the body and the left leg. This action begins with turning in the rear knee and rotating on both feet toward the pitcher. The exact pathway followed by the bat is determined by the batter when reacting to the flight path of the pitch (see figure 13.5b-13.5c).	The batter initiates the sequential rotation of the legs, hips, and torso from the ground up. More massive body segments (i.e., legs, hips, torso) are vigorously rotated and then suddenly decelerated. Their angular velocity is multiplied up through the batter's body, along the length of the batter's arms, and out to the barrel section of the bat. The bat rotates in flail-like fashion around an axis formed by the left shoulder and the batter's wrists.
When the bat is swung around at high velocity, the batter is forced to lean backward away from the bat.	The tremendous angular velocity of the bat requires the batter to generate considerable centripetal force. The batter leans away from the bat and in doing so combines lean with the downward pull of gravity on the body mass to counterbalance the inertial and centrifugal pull of the bat.
The batter attempts to strike the ball at the bat's "sweet spot." To drive the ball a great distance, the batter tries to avoid a slicing impact, aiming instead to hit the ball through its center.	The batter strives to strike the ball at the bat's center of percussion (i.e., the bat's sweet spot). If the force applied by the bat fails to pass through the ball's center of gravity, the ball can be undercut and given backspin. A Magnus effect is applied to the ball, counteracting gravity and giving it lift. The ball hangs up in the air, making it an easy target for a fielder. An overcut gives the ball topspin. In this case the Magnus effect joins the downward pull of gravity, so the ball arcs quickly down toward the earth.
The batter's arms swing forward and away from the chest. The left arm extends as the bat swings around the body into the follow-through. The batter rotates around the left leg and the left side of the body (see figure 13.5d). Vision is on the flight of the ball.	The batter's body rotates around the axis of the left side of the body. This action extends the radius of rotation from the left shoulder out to the barrel of the bat. A follow-through completes the striking action and dissipates the momentum built up in the rotary action of the batter's body and the bat.
The flight of the ball depends on what occurs when the ball is hit and the effect of environmental conditions.	The velocity, direction, and distance of ball travel after the ball is hit depend on a large number of factors, including the following: • The momentum of the ball at the instant of impact, the momentum of the bat at the instant of impact, the elasticity (recoil) of the ball • The direction in which the bat and ball are moving at the instant of impact, the point of impact between the bat and the ball • The spin put on the ball after impact • Environmental conditions such as altitude, temperature, humidity, and airflow

Swinging and Rotating

The highlights of swinging and rotating, as in a back giant on the high bar in gymnastics, shown in figure 13.6, are as follows:

• The manipulation of the body segments directly influences the mechanical principles that control angular motion and govern swinging. In swinging skills, an athlete stretches (or lengthens) out on the downswing. This action moves the athlete's center of gravity as far from the axis of rotation as possible. By carrying out this action, the athlete allows gravity to exert maximal accelerative torque to the body on the downswing.

• To rise high on the upswing, an athlete counteracts the decelerating effects of gravity by flexing at the hips and shoulders. This action pulls the athlete's center of gravity closer to the axis of rotation, which reduces the decelerative torque of gravity. The timing of changes in body position directly influence the angular velocity when swinging and rotating.

• During performance of somersaults and other rotary skills in the air, the use of a lever arm and turning force is required (i.e., a force that does not pass through the athlete's axis of rotation is applied to the athlete's body at the start of the skill). By applying a force that is offset from the athlete's axis of rotation, a torque is applied to the athlete's body. When this torque is greater than the resistance, the athlete will rotate. A slight body lean in the direction of rotation at the instant of takeoff is a common method for applying this turning or rotating force.

• Reducing the athlete's rotary resistance by altering body segment length and subsequently the radius of gyration during flight increases the athlete's angular velocity (i.e., the rate of spin). Rotary inertia is reduced as the athlete uses his muscles to flex his body segments toward the axis of rotation, shortening the radius of gyration. Angular velocity is decreased by extension of the athlete's body

mass outward from the axis of rotation. This action increases the athlete's rotary inertia and reduces the athlete's rate of spin.

AT A GLANCE

Movement of an Implement

• To move an external implement effectively, the athlete needs to coordinate the movement of his anatomical segments (kinetic link) before transferring this velocity to the implement.

• The ability to relax the antagonist muscle groups and contract the agonist muscles in the correct time and sequence can make each body segment move faster and produce a more accurate swing pattern.

• Altering the length of body segments (radius of gyration) can work with or against gravity to vary the rate of angular velocity.

FIGURE 13.6   Sagittal plane view for back giant swing on the high bar.

Adapted from M. Adrian and J. Cooper, *The Biomechanics of Human Movement* (Indianapolis, IN: Benchmark Press, 1989), 698, by permission of The McGraw-Hill Companies, Inc.

Swinging and Rotating Characteristics and Mechanical Principles

Technique	Mechanics
A gymnast frequently begins in a front support position and "casts" (i.e., thrusts against the bar) to elevate the body to a position where the center of gravity is well above the bar.	By elevating their center of gravity high above the bar, gymnasts increase their potential energy. Gravity then has the opportunity to accelerate them from greater height on the downswing. The higher that gymnasts raise their center of gravity, the greater the time is during which gravity can accelerate them downward. This action increases their angular velocity and angular momentum.
The handgrip for a back (i.e., backward) giant is an overgrip. For a front (i.e., forward) giant, the handgrip is an undergrip (a reverse grip).	As the athlete stands facing the bar, an overgrip has the knuckles above the bar and thumbs below. In a back giant, the thumb "leads" the athlete through the skill, and the bar "rotates" into the hand for a strong, safe grip. An undergrip, or reverse grip, is used for a front giant. As the athlete stands facing the bar, the hands are turned out so that the thumbs are above the bar and the knuckles below. Again, the thumb "leads" the athlete through the skill and the bar "rotates" into the hand for a strong, safe grip. Failure to use the correct grip will cause the athlete to come off the bar during the performance of either of these skills. For greater safety with the advanced high-bar skills, gymnasts use handgrips with dowels inserted into them, which help lock the hand on the bar.
Once well above the bar and beginning the descent, the gymnast stretches to full extension (see figure 13.6a).	By stretching to full extension, gymnasts push their center of gravity as far as possible from the axis of rotation (i.e., the bar). This position allows gravity to apply the greatest amount of accelerative torque to the gymnast's body on the downswing. Just before reaching a position directly below the bar, the gymnast flexes the muscles of the spine. This action prestretches the abdominal muscles, which are ready to flex the hips when the athlete rises on the upswing (see figure 13.6b).
Gravity accelerates the gymnast's body downward.	Gravity applies an accelerative torque to the gymnast's body. This torque increases until it is maximal when the gymnast is stretched horizontally from the bar. From that point, although the gymnast's angular velocity and momentum continue to increase, the amount of torque applied to the gymnast's body gets progressively less until it is zero when the gymnast passes directly below the bar.
After passing below the bar, the gymnast flexes at the hips and at the shoulders (see figure 13.6c-13.6d). Beginners may also flex at the knees.	Flexion at the hips and shoulders brings the gymnast's center of gravity closer to the axis of rotation. The decelerative torque of gravity is reduced. By flexing, gymnasts also pull their mass closer to the bar, which reduces their rotary resistance (i.e., rotary inertia).
In a flexed body position, the gymnast rises upward. The flexion in the body is progressively eliminated as the athlete moves to a position vertically above the bar (see figure 13.6e-13.6f). Some pulling action is applied to the bar to get the gymnast's body over the top of the bar.	With a reduction in the gymnast's rotary inertia and a reduction in gravity's decelerative torque, the gymnast is able to rise to a position vertically above the axis of rotation (i.e., the bar). Immediately before reaching the vertical position, the athlete can extend the body to slow (i.e., control) the rate of spin around the bar.
Once above the bar and beginning the descent, the gymnast extends the body again to repeat the process described in the preceding steps.	Above the bar, the gymnast straightens out to shift the center of gravity as far from the bar as possible. This action allows gravity to reapply maximum torque to the gymnast's body on the downswing. In this way, continuous giants are performed.

Weightlifting

These are the highlights of weightlifting, as in the clean and jerk, shown in figure 13.7:

• Sports that use pull–push actions such as powerlifting and weightlifting require the athlete to apply force continuously throughout the desired range of movement. In weightlifting (or powerlifting), athletes need to generate a tremendous amount of force to overcome the inertia of the weight, usually only once in a competition (one repetition or 1RM). In training, however, they perform several repetitions, depending on the training phase (for example, 8RM).

- Regardless of the number of repetitions, the principle is the same: The athlete wants to apply maximum force to a heavy resistance to overcome the inertia of the heavy object. To do this, the athlete simultaneously uses the largest number of body segments that can be applied to the task (e.g., the legs, back, chest, shoulders, and arms). This simultaneous action distinguishes weightlifting from high-velocity throwing, kicking, and striking skills, in which a sequential movement of body segments occurs.

- The two Olympic weightlifting events are the clean and jerk and the snatch. The clean and jerk is used to hoist the heaviest weight and is a two-phase lift. The bar is first pulled to the chest, where a pause occurs. The athlete then jerks (pushes) the bar to arms' length above the head. The snatch differs from the clean and jerk in that it is performed with a continuous pulling action with no pause at the chest (Kompf and Arandjelović 2017). In Paralympic powerlifting the athletes compete in a controlled bench press, because they typically have lost lower-limb function and therefore cannot stand for the clean and jerk or the snatch.

- The clean and jerk, the snatch, and the bench press are power events in which coordinated strength and speed are required. In these events the athlete applies great force in a limited period (impulse) to accelerate the barbell upward. At the height of the barbell's upward movement, the athlete must move at high speed into a position where the barbell is stabilized at the chest (for the clean) or above the head (for the jerk and the snatch).

a   *b*   *c*   *d*   *e*

FIGURE 13.7   Frontal plane view of the clean and jerk in weightlifting.

Weightlifting Characteristics and Mechanical Principles

Technique	Mechanics
The clean and jerk uses two types of leg action. With a heavy barbell, an athlete cleans the bar by first pulling on it and lifting it as high as possible. Without pause, the athlete either squats under the bar or splits the legs forward and backward. The athlete then rises up out of this position, with the bar held at the chest ready for the jerk. To complete the jerk, an upward jerking action is combined with a lunging motion in which the legs are split forward and backward.	The faster the bar moves upward and the higher it travels, the more time that is available for the athlete to perform the squat or leg split. The more powerful the athlete is or the lighter the resistance is, the faster and higher upward the bar will move. A bar pulled upward to a high position does not require high-speed performance of a deep squat or a leg split as the athlete moves under the bar.
The starting position in the clean has the athlete in a shoulder-width stance with the bar in front of the shins. The athlete grips the bar with the hands slightly wider than shoulder-width and equidistant from the plates. The arms and back are straight, and the legs are flexed no more than necessary. The athlete's back is angled at 45° to the horizontal (see figure 13.7a).	The starting position for the clean is such that the bar will be pulled predominantly in a vertical manner. The athlete's partially flexed legs (i.e., approximately a half squat) are in a mechanically efficient position for a powerful extension. The arms are extended and immediately transmit force from the athlete's legs and back to the bar.

> *continued*

Weightlifting Characteristics and Mechanical Principles > *continued*

Technique	Mechanics
The clean begins with a powerful extension of the athlete's legs and back to accelerate the bar upward.	The resting inertia of the bar is overcome by a powerful extension of the athlete's legs and back. The legs and back extend simultaneously, transferring the pull to the bar through the arms. The bar rises in a vertical direction.
When the legs have completed their extension, the athlete continues to pull upward by flexing the arms. The athlete rises up onto the toes and hyperextends the back. The pull of the arms completes the vertical movement of the bar, which travels upward close to the athlete's body. The head is thrown back (see figure 13.7*b*).	The large muscles of the legs and back apply the greatest force to the bar. Driving up onto the toes and pulling upward with the arms allow the athlete to continue applying force to the bar over a long period. Athletes pull the bar upward as close to their line of gravity as possible. In this way the center of gravity of the bar stays above (and within) the athlete's supporting base, and the athlete and bar are optimally stabilized.
When the barbell has risen to just below the pectoral muscles, the athlete squats down under the bar, simultaneously rotating the arms forward so that the bar is pulled in toward the upper chest and held in place with the arms (see figure 13.7*c*).	The greater the mass (weight) of the barbell is, the more critical are the upward pull on the bar and the velocity with which the athlete rotates the arms forward and squats under the bar. A sufficiently high pull, fast arm rotation, and equally fast squatting action are essential.
The athlete uses leg strength to lift up from a front squat position (see figure 13.7*c*-13.7*d*).	With the bar at the chest and in a squat position, athletes must use great leg strength to battle the deadweight (i.e., inertia) of the barbell (plus the inertia of their own body mass) so that they can rise to a standing position. With insufficient strength or too massive (i.e., too heavy) a barbell, the athlete will either fail to get up out of the squat or be physically drained from the effort of standing up and fail to complete the jerk.
In a standing position and with the bar held at the chest, the athlete dips and flexes the legs slightly. The weight discs at either end of the bar cause the bar to flex downward and then upward.	A slight flexion prestretches the leg muscles before the extension of the legs. The upward recoil of the bar is timed to coincide with the leg extension to assist in thrusting the barbell upward.
The instant the athlete's legs have completed their thrust and the bar is rising upward, the legs split forward and backward. One leg is extended directly backward at an angle of about 45°. The opposite foot steps forward 25 to 28 cm (10 to 11 in.). This action allows the athlete to lunge forward under the bar (see figure 13.7*e*).	The athlete's leg thrust and arm extension (with no leg split) is adequate to elevate a lightweight barbell. With a heavier resistance (which cannot be elevated to arm's length above the head with leg thrust and arm extension alone), the athlete is forced to split the legs at high speed and drop very low under the bar.
With the bar held above the head at arm's length, and with the legs split forward and backward, the athlete carefully brings both legs back toward their original position. The athlete finishes the lift standing with the feet shoulder-width apart and with the barbell held at arm's length above the head. For approval from the judges, the athlete must demonstrate control over the barbell for 3 s.	With the barbell above the head, the combined center of gravity of athlete and barbell is raised upward, and with increased height, both athlete and barbell become progressively unstable. The athlete must struggle to keep the line of gravity of the barbell centralized above a small, narrow base to avoid losing control and failing in the lift.

Combat Throwing

The highlights of combat throwing, such as the hip throw in judo, as shown in figure 13.8, are as follows:

- Athletes in combat sports such as judo use combinations of rotation, pulling, pushing, and lifting to lessen an opponent's stability and set up the opponent for a throw. Opponents counter an attack by adjusting their center of gravity, with respect to their base of support, by leaning toward or away from their opponent.

- To increase stability and make themselves less vulnerable, athletes widen their base and lower and centralize their center of gravity. Weight divisions in combat sports are intended to provide an even playing field, with respect to body mass of the athlete.

- Maintaining stability and destroying the stability of an opponent in judo, and other combat sports, are predominantly a battle of one torque versus another. Judo involves pushing, pulling, lifting, and rotating, all of which are designed to maintain the attacker's stability while disrupting that of the opponent. Precise timing, body segment coordination, and superfast muscle reactions are essential.

- The manipulation of the line of force can create the force lever to push, pull, lift, and spin an opponent around axes formed by the athlete's feet, hips, back, and shoulders. Leg sweeps in judo are a common method of eliminating an opponent's base of support. The intent of the attacker is to shift the opponent's center of gravity outside its supporting base and thereby destroy the opponent's stability. Superfast combinations of pushing, pulling, lifting, and rotating are used to achieve this end.

FIGURE 13.8   Sagittal plane view of the hip throw in judo.

Combat Throwing Characteristics and Mechanical Principles

Technique	Mechanics
Judoka (i.e., judo practitioners) face each other in a standing position with their bodies slightly lowered and their legs partially flexed. Their feet are at right angles to each other and shoulder-width apart. They take fast, shuffling, flat-footed steps, and the athlete's body weight is frequently positioned closer to the front foot.	Lowering the center of gravity increases the stability of the athlete. Positioning the feet at right angles gives good stability side to side and front to back. Fast, shuffling steps increase stability because they limit time spent on one foot. Having the line of the center of gravity closer to the front foot readies the rear leg for leg sweeps and other destabilizing actions used against the opponent.
Judoka begin their attacking and defensive maneuvers by grasping each other's tunics on the collar at shoulder level with one hand and at the sleeve with the other (see figure 13.8a). Grips are suddenly altered according to the throw being attempted.	Grips on the tunic are designed to facilitate pull–push and rotational movements. Any combination can be used. Applying force at the sleeve rotates opponents around their long axis. Applying force at the collar and upper lapel produces forward and backward movement or rotation.
With grips on each other's tunics, judoka circle each other, waiting for the opportunity to initiate a throw.	Preparation for a throw is a series of pull–push rotary actions in which the prime objective is to move the opponent's center of gravity into a position of minimal stability.

> continued

Combat Throwing Characteristics and Mechanical Principles   > *continued*

Technique	Mechanics
In a hip throw, which incorporates a lifting action, the attacker grasps the rear part of the collar with the right hand. The left hand grips the opponent's sleeve below the right arm. The attacker pulls the opponent forward with both hands (see figure 13.8a).	Pulling the opponent's body forward, the attacker lessens the opponent's stability by moving the line of the opponent's center of gravity closer to or beyond the edge of its supporting base. A high grip on the rear collar maximizes the force arm from the collar to the axis of the attacker's hips, over which the opponent will subsequently be rotated. The grip below the right arm will spin (i.e., rotate) the opponent around his long axis.
The moment the opponent is successfully drawn forward and destabilized, the attacker quickly initiates a pivot on his left foot by stepping across with his right foot while flexing the legs (see figure 13.8b).	By rotating his body to the left, the attacker prepares to use his hip as an axis of rotation over which the opponent will be pulled. Flexing his legs, the attacker not only increases his own stability but also prepares to destabilize the opponent by lifting him out of contact with the mat (and the earth) when he extends his legs.
The attacker's lower back is pressed against the opponent's thighs. The attacker's upper back is pressed against the opponent's abdomen. The opponent's upper body is pulled downward over the attacker's hip with both hands, while the opponent's feet are lifted off the mat when the attacker straightens his legs (see figure 13.8c).	The attacker's upper hip becomes an axis of rotation over which the opponent rotates. A force is applied by the attacker's hands, which pull (and rotate) the opponent's upper body downward. Another force is applied by the extension of the attacker's legs, which drive the opponent's lower body upward. The opponent spins around the axis of the attacker's hips.
The opponent's lower body is forced upward, taking his feet out of contact with the mat. The opponent's upper body is pulled downward. The opponent rolls over the attacker's hip (see figure 13.8c).	The attacker's downward pull and leg extension eliminate the opponent's contact with the mat. Frictional forces between the opponent and the mat no longer exist. The opponent is now defenseless and will be thrown onto his back.
A breakfall is performed by the opponent on contact with the mat (see figure 13.8d).	Totally destabilized, the opponent prepares for the impact with the mat that will occur after the fall. The opponent correctly enlarges the time and area over which the impact is made with the mat.

Tackle in Football

The following are the highlights of the tackle in football (Kempton et al. 2014), as shown in figure 13.9:

• Because of the high impact forces in the football tackle, the first component of an effective football tackle is for the athlete to get into a safe position before the collision between the two players. The most important safety concern is to protect the head of the athlete. Achieving this goal requires the player to keep his head on the outside of the tackle and ensure that his head does not come between the opposing player and the ground.

• The tackle requires slowing down, stopping, and then (if possible) pushing the opposing player backward and toward the ground. To arrest this motion requires effective manipulation of the forces, energy, and momentum present in a collision when two or more objects come together (Hendricks et al. 2014).

• Athletes need to establish a suitable base of support before contact with the opposing player,

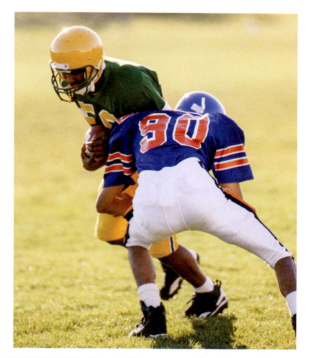

FIGURE 13.9   Sagittal plane view of tackling in football.
Photodisc/Getty Images

because this stance will enable an efficient transfer of forces from the solid ground, through the lower limbs, then the torso, and finally the shoulders and arms as they make contact with the opposition. The proper coordination of the lower and upper limbs is essential for a safe and effective tackle.

• To arrest their own motion safely, athletes apply a stopping force to their bodies over as large a distance and time as possible. In addition, they spread the impact over as large an area as possible. In this way the pressure applied at any one spot to the athlete's body is reduced. Flexing the legs and rolling during landing, as well as using padding and crash pads, gradually dissipate the forces that occur during impact. An athlete can then avoid injury (which occurs when great force is stopped in an instant in a small area) and maintain control and stability when the impact occurs.

Tackle in Football Characteristics and Mechanical Principles

Technique	Mechanics
The football tackle requires the tackling player to run toward the opposing player and to place himself within arm's length of the opposition. Because both players are moving, getting into this position requires advanced skill to generate a safe and effective tackle.	To transfer the inertial forces effectively, the tackling player must make contact with the opposition. Getting into a safe position with the head well clear of the impact zone is paramount. Doing this requires a technically correct combination of position and acceleration to put the tackling player into the desired position.
After the tackling player is within reach of the opposing player, he needs to develop a stable base of support to tackle an opposing player effectively in the National Football League, Rugby League, or Rugby Union.	The stable base of support is generated with placement of the feet as far apart as possible. This stance allows a greater range of movement of the line of gravity within the base of support. The ability to generate a tackling force can also be enhanced if the base of support is greater in the direction of travel, because this stance also allows greater torque to be produced.
The first point of contact is a large and stable segment of the upper torso, ideally the shoulder girdle. After body contact is made with the shoulder, the arms are used to wrap around the opposing player to contain and direct his position.	Because of the high impact forces in the football tackle, the first point of contact must be made with a stable and supported musculoskeletal joint, namely the shoulder girdle, rather than with the arms; the latter anatomical limbs are more susceptible to broken bones because they are smaller and have less muscle around the segment.
To ensure the safety of the tackling player, extreme care is required in the positioning of the head. To avoid the extreme impact forces that are generated between the two players as they fall to the ground, tackling players must ensure safe tackling technique with the head away and on the outside of the opposing player.	The impact forces generated within a tackle can cause high velocities, and when the fast-moving anatomical segment makes a rapid stop as it hits the ground, the extremely high deceleration can be greater than the safe acceleration of the musculoskeletal system of the human allows, resulting in severe damage.
After contact is made between the tackling player and the opposition, the objective of both players is to continue moving in the original direction. Depending on the forces generated by the two players, the result will be that either (a) the tackling player pushes the opposing player backward, (b) the opposing player continues in the original direction and pushes the tackling player backward, or (c) the forces generated by the two players are equal and at impact the two players remain stationary.	The final direction following the tackle collision depends on the resultant force generated by each player. Because the two players are traveling in different directions, the net resultant force is the sum of their original inertia before impact.

> *continued*

Tackle in Football Characteristics and Mechanical Principles   > *continued*

Technique	Mechanics
When falling to the ground, the players need to prepare for impact with the ground by slightly bending each anatomical joint to provide for absorption of the impact. When the players hit the ground, their velocities decrease to zero.	The fall to the ground can occur from various heights (e.g., from as high as an opponent's shoulders). Falls also occur at various velocities. The opponent can add his own muscular force to that of gravity. The momentum generated during the fall equals the mass of the falling player multiplied by his velocity. The player's velocity (and momentum), which built up during the fall, is reduced to zero on contact with the ground. The force with which the player hits the ground is applied against the athlete's body by the reaction of the earth.
Safe landing on the ground is achieved if the athlete's arms and legs simultaneously make contact with the ground. In this way vital organs within the trunk are protected because the trunk does not absorb the full force of the fall.	In striking the ground with the arms and legs, players use their shoulders, thighs, and lower leg as shock absorbers. By carrying out this action, players extend the time during which force is applied against the earth and, in reaction from the earth, against the athlete. An extended time over which force is applied causes a more gradual change in the player's momentum (and kinetic energy), reducing the chance of injury.
By rolling and contacting the ground with extended arms and legs, the player enlarges the surface area of the body that contacts the ground.	Enlarging the contact area of the body as the player's body contacts the ground decreases the pressure applied at any point on the athlete's body. Enlarging the contact area and extending the time over which force is applied to the earth (and from the earth to the athlete) significantly reduce the possibility of injury because of the tackle. A situation in which the player falls vertically onto the ground brings an athlete to a sudden stop. Considerable force is applied to the earth, and from the earth against the athlete, in a short time. Flexing the legs, squatting down, and rolling extend the time during which force is applied by the player against the ground and, in reaction, by the earth through the ground to the player.

Swimming

The highlights of swimming, as in the freestyle stroke, shown in figure 13.10, are as follows:

• When moving through the fluid of water the two key aspects are to increase propulsion and reduce resistance. The propulsion of the athlete (swimmer) as she moves through the water depends on the product of stroke length and stroke rate. Stroke length refers to the distance over which the swimmer pulls and pushes with each stroke. Stroke rate refers to the number of strokes per minute. The resistance of the athlete depends on the fluid frictional forces and the drag of the swimmer.

• In the freestyle, back crawl, and butterfly, the swimmer's hands and forearms produce the greatest propulsive forces. In these strokes the legs also contribute to propulsion when kicking, and the swimmer's resistance decreases because her kick can place her body closer to horizontal, which minimizes form drag (Formosa et al. 2014).

In the breaststroke, both arms and legs provide similar powerful propulsive forces depending on the individual's technique.

• Swimmers need to overcome several frictional forces: (1) form drag, caused by the swimmer's shape and body position in the water; (2) surface drag (skin friction), caused by the water rubbing against the swimmer's body surfaces; and (3) wave drag, caused by the resistance of the water piling up in front of the swimmer. A horizontal body position in the water combined with smooth surfaces (i.e., resulting from shaving and the use of slick swimsuit fabrics) helps to reduce form and surface drag. Swimmers reduce wave drag and attempt to be energy efficient by controlling up-and-down body motions and avoiding thrashing, splashing arm and leg actions.

• Swimmers emphasize stretching and reaching forward to increase the length and range of the stroke as well as maintaining a high stroke rate. Swimmers aim for rhythmic, relaxed swimming and try to develop a feel for the water.

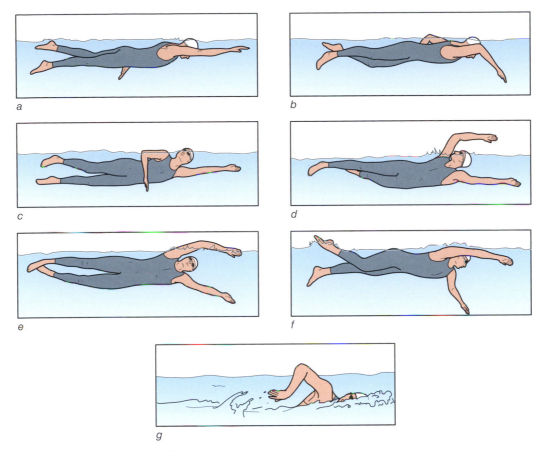

FIGURE 13.10   Sagittal plane view of freestyle swimming stroke.

Swimming Characteristics and Mechanical Principles

Technique	Mechanics
The freestyle (sometimes called Australian crawl) is the fastest of the four swimming strokes. Arm action, leg kick, body position, and breathing are synchronized and contribute to the velocity of the swimmer through the water.	In the freestyle each arm has a propulsion and recovery phase, and depending on the swimming style this may overlap or be independent (see figure 13.10c). Both freestyle and backstroke use this cyclic rotary action, whereas in the breaststroke and butterfly strokes the two arms pull, push, and recover simultaneously. In the freestyle the swimmer's hands and forearms produce most of the propulsive force.
The swimmer's body lies horizontal in the water. The chest is pressed down, and the face is immersed (see figure 13.10a-13.10b). Swimmers often look at the bottom of the pool to raise the legs. The head, shoulders, and hips all roll to the same side to enable breathing (see figure 13.10c). The leg kick occurs just below the surface (see figure 13.10a-13.10b). During the arm recovery, the arm is flexed at the elbow (see figure 13.10g). Stretching forward with the leading arm and rolling along the long axis of the body are coupled with breathing and the recovery of the propulsive arm (see figure 13.10c-13.10d).	Keeping the head down maintains good body alignment and reduces form drag in the water. It also enhances the ability to glide with each stroke. (Raising the head causes the legs to drop, which increases wave, form, and surface drag.) If the arm recovery is swung across the swimmer's long axis, the swimmer's body will "snake" through the water, increasing both form and surface drag. A flexed arm during recovery reduces its rotary resistance and makes it easier for the muscles involved to bring the arm forward. Slamming the arm and hand into the water at entry causes the body to react in the opposing direction and bounce up and down. This action increases wave, surface, and form drag. Rolling along the long axis presents a narrow profile to the water, which minimizes drag.

> continued

Technique	Mechanics
Entry of the hand is made in front of the head in a position in line with the shoulder (i.e., at shoulder-width). At entry, the arm flexes slightly, and the elbow is positioned above the hand. The fingers slice into the water before any other part of the hand and arm, and the swimmer attempts to "catch" or "fix" the pulling surfaces of the hand and forearm into the still, nonturbulent water ahead and to draw the body past this catch, or fixed, point (see figure 13.10e-13.10f).	At the entry of the hand and forearm into the water, lift forces generated by the flow of water from the fingertips up the hand and arm aid partially in propulsion. The swimmer uses momentum to move past the catch position. Still, nonturbulent water provides an area of water in which the swimmer can pull and push more effectively. This form of drag propulsion is greatest when the hand and forearm pull and push backward parallel to the direction of swim.
As the body moves forward past the catch, the pulling arm flexes at approximately 90° with the elbow high and the hand positioned lower in the water (see figure 13.10b-13.10c). The degree of flexion at the elbow varies throughout the arm's pull–push phase. The pull–push of the propulsive arm is timed to occur as the opposing arm is recovered and thrust forward in the water (see figure 13.10d). The propulsive arm exits the water at hip level (see figure 13.10b-13.10c).	Partial flexion of the propulsive arm at the elbow reduces its lever length relative to the shoulder joint but does not reduce the active surface area of the swimmer's arm as it pulls through the water. Flexion at the elbow also makes it easier for the swimmer's muscles to pull the hand and forearm along their propulsive pathway. The pull–push with the hand and forearm is completed at hip level and optimizes the impulse (i.e., force × time) over which the propulsive force is applied. The resultant of lift and drag acts also as a propulsive force predominantly at the beginning and end of the pull–push action.
Some lateral and vertical motion of the propulsive hand and arm occurs during the pull and push. The swimmer rolls along the long axis of the body as the leading arm is stretched forward. Rolling assists with breathing and with recovery of the propulsive arm after it has completed its pull–push action (see figure 13.10c-13.10e).	Anatomical limitations in the shoulder and elbow joint, coupled with reactions to head movements and arm recovery, cause some lateral and vertical movement of the propulsive hand and arm as they move backward through the water. The angle of attack of the hand varies continuously throughout the stroke to maximize lift and drag propulsion.
The flutter kick has an arc of movement of about 0.3 to 0.6 m (1 to 2 ft) in size and occurs just below the surface of the water (see figure 13.10a-13.10f). It has a downbeat–upbeat action and some rolling lateral motion, which is caused by the rolling of the body itself. The kick starts at the hip and works down to the ankle in a whiplash motion. The flutter kick is synchronized with breathing and the arm action.	The flutter kick cannot compare with the hands and arms in terms of propulsion, but it plays an important role in holding the swimmer's body in a streamlined, horizontal, drag-reducing position. Much depends on the swimmer's having great mobility and flexibility in the ankles. Kicking too deep increases drag without contributing much to propulsion. The depth of the kick depends on the swimmer's build, strength, and stroke rate. A stiff-legged flutter kick consumes too much energy and is incorrect.
Breathing depends on the stroke pattern developed by the swimmer. Many swimmers breathe every stroke cycle, and breathing every two strokes is a common pattern in races that are longer than a single length (50 m).	Breathing should not in any way hinder efficient propulsive technique. Lifting the head to breathe causes the legs to drop, increasing form drag.
The athlete tries to swim with a smooth, rhythmic, nonjerky, cyclic action. Up–down, side-to-side motions are eliminated with the torso held fairly rigid. The athlete's arm action is long, rangy, and precise and gives an impression of relaxation even at high speed.	Jerky, up–down, and side-to-side motions increase the resistance of the water, in particular wave drag (which increases according to the cube of the velocity). An increase in wave drag demands an immense increase in energy expenditure from the athlete and generates phenomenal resistance the faster the athlete swims. The athlete should aim for fluid, relaxed motions.

Technique	Mechanics
Swimmers frequently shave the skin exposed to water flow and use bodysuits that are patterned with vortex generators. Dolphin kicks are used for several strokes immediately after entrance into the water from the start and also when the swimmer comes off the wall after completing a turn. The rules of competitive swimming limit the distance that can be swum underwater.	Shaving reduces surface drag. Vortex generators are lines of small bumps or specifically located seams on bodysuits. The objective of vortex generators is to reduce both form and surface drag caused by the swimmer both on and below the surface of the water. Because bodysuits increase buoyancy, the swimmer is more easily able to maintain a horizontal position in the water, which dramatically reduces form drag.

Kicking a Ball

The highlights of kicking a ball, as in a punt in football, shown in figure 13.11, are the following:

• When kicking a ball through the fluid of air, the athlete needs to coordinate the segment movements (kinetic link) for contact with the ball. The high-speed movement of the lower limb requires the athlete to generate exceptional velocity of the foot. This flail-like action begins with the slower motions produced in the athlete's longer, larger, and heavier limbs and then progressively becomes faster and quicker as the athlete passes the motion on to lighter, less massive body parts, ultimately being transferred at impact with the ball.

• Kicking effectively requires a controlled and rapid acceleration of the athlete's body segments, beginning with a stable foot placement to provide stability before leg swing. Then, starting with the upper torso, the athlete progressively accelerates the torso, hips, thigh, and shank and finally the foot before contact with the ball. To achieve the greatest possible velocity, antagonist muscle groups must be completely relaxed and agonist muscles must contract in sequence, helping to make each body segment move faster than the previous one.

• Kicking skills differ from throwing and striking skills in that the last and fastest moving segments or links in the whip-like sequence are the athlete's lower leg and kicking foot. The progressive acceleration of body segments in kicking skills is similar to that used for throwing and striking skills.

• Depending on the desired outcome, the impact of the kick can be off center to create spin on the ball. In the right conditions this spin can then alter the path of the ball (Magnus effect), causing the ball to follow a curved path.

• Each movement pattern in kicking skills (and throwing and striking skills) contains some preparatory actions, commonly called approach, setup, and windup or backswing. These actions help place the body in relationship to the ball to provide optimal generation of velocity and ultimately the application of force. A backswing of the leg provides additional distance over which the leg is accelerated. The backswing prestretches the kicker's muscles in preparation for their explosive recoil. Relaxation and flexibility help produce an optimal windup.

| a | b | c | d | e |

FIGURE 13.11 Frontal plane view of punting (kicking) a football.

Kicking a Ball Characteristics and Mechanical Principles

Technique	Mechanics
Kicking is a striking action used to apply force with the foot. Punting a football uses the kicking leg in much the same way that a javelin thrower uses the throwing arm or a batter uses both arms and a bat to strike a baseball.	Punting relates mechanically to striking actions made with a club or bat. It also relates mechanically to throwing actions in which the object thrown is lightweight and moves at high velocity. Kicking also requires the athlete to simulate the actions of a whip. Long legs, great muscular strength, and a huge range of movement help generate tremendous impact force when the kicking foot contacts the ball.
An athlete punts a football by stepping forward onto the supporting foot. This foot is positioned in front of the athlete's body, which is inclined backward (see figure 13.11a). The ball is positioned so that it contacts an extended kicking leg. Vision is on the ball, and the arms are extended from the body to counterbalance the powerful swing of the kicking leg.	In punting, athletes overcome the inertia of their body (and that of the ball) by moving forward toward the direction of kick. This movement gives momentum to both the athlete and the ball. The supporting foot is placed well ahead of the athlete's center of gravity so that the athlete can move forward, up, over, and beyond this foot, which acts as an axis. The athlete's forward movement extends the time over which force is applied to the ball. The outstretched arm assists the athlete in maintaining stability throughout the kicking action.
As the athlete steps into the kicking stance, the kicking leg (which is flexed at the knee) trails well to the rear of the athlete's body (see figure 13.11a).	The trailing action of the kicking leg and the backward tilt of the athlete's body extend the time and distance over which force is applied to the ball. The kicking foot travels over a huge arc from its backswing position to the point where it contacts the ball.
As the thigh of the kicking leg is swung forward, the lower leg trails to the rear and temporarily rotates in the opposing direction. A 90°angle occurs at the knee (see figure 13.11b).	The forward swing of the thigh and the backward motion of the lower leg simulate the initial action of a whip, in which the more massive lower part of the whip moves forward while lighter sections momentarily move in the opposing direction.
After accelerating forward and upward, the thigh of the kicking leg slows down. The lower leg and kicking foot now accelerate. The athlete will have positioned the ball ahead of the body so that the kicking foot meets it while moving at high velocity.	The high angular velocity of the thigh is arrested. This causes its angular velocity to be passed on to the lower leg, which is less massive and has less rotary resistance. The angular velocity of the lower leg is dramatically increased. The ball is positioned so that the impact occurs when the kicking foot has maximum momentum.
The kicking leg is extended when it contacts the ball (see figure 13.11c).	A powerful extension of the kicking leg occurs as the lower leg rotates around the axis of the knee joint. The kicking foot, simulating the tip of the whip, now moves at maximum velocity.
The athlete rises up on an extended supporting leg. The athlete's body tilts backward away from the direction of the punt (see figure 13.11d-13.11e).	Rising onto the toes of an extended supporting leg, coupled with a backward body tilt, enlarges the arc through which the kicking leg swings and through which force is applied to the ball.
The athlete's body moves forward past the supporting foot and in the direction of the kick. After the foot contacts the ball, the kicking leg continues with a follow-through. Flexion reoccurs at the knee of the kicking leg (see figure 13.11e).	Forward movement of the athlete's body continues the application of force to the ball. Maximum angular velocity of the lower part of the kicking leg occurs just before contact with the ball. Partial flexion of the kicking leg allows the foot to swing upward through a large arc to a position above the athlete's head. The arms are often extended sideways to maintain stability.
The distance the ball travels depends on how much force is applied to the ball, the trajectory angle of the ball, environmental conditions, and whether the ball spirals around its long axis or tumbles end over end.	The velocity of the kicking foot at the instant it contacts the ball, coupled with the angle of release, determines the distance that the ball travels. Feeding the ball onto the kicking foot so it has a high trajectory angle when contact is made is appropriate when the punter is kicking with a following wind. A low trajectory angle is used for kicking against the wind.
For a spiraling kick, the athlete positions the ball so that its long axis is slightly offset from the intended direction of the kick. The kicking foot is swung directly ahead and is drawn across the long axis of the ball.	Contacting the ball so that the kicking foot moves across the ball's long axis applies torque to the ball, causing it to spin. A spiraling ball has greater stability and travels a greater distance than a ball that tumbles end over end.
The distance the ball travels in flight will be affected by environmental conditions.	Wind direction and altitude (air resistance) have a considerable effect on the distance that the ball travels.

Paddling a Watercraft

These are the highlights of paddling a watercraft, as shown in figure 13.12:

- When paddling a watercraft (such as a rowing skull, ski, or kayak), the athlete uses a sequential movement of body segments to achieve propulsion. The paddling stroke typically has four steps. The first is the catch, when the oar (or paddle) is put into the water. The athlete's legs bend slightly during a forward lean of the torso. The head is kept up, and the blade of the oar is lowered to the surface. Pushing is unnecessary because the oar is balanced to enter the water.

- The second phase is the drive-power portion of the stroke, that is, pulling the oar through the water. The blade is set so that it enters the water perpendicular to the surface. The push begins with the legs, followed by the back muscles (McKean and Burkett 2014). Finally, the athlete pulls with the arms, moving the oar up to the torso just below the chest. Pulling through the water at a deeper angle doesn't provide any extra power. The blade should move just below the surface of the water. Athletes who are sloppy and don't dip the oar in far enough can "catch a crab," scooping up a splash of water as the oar flies upward. On a multiperson racing shell, this action can rock the entire boat, throwing off the rhythm of the crew. To reduce the resistance of the craft as it moves through the water, horizontal or vertical movements need to be reduced as much as possible.

- The third phase is the finish stage of the underwater stroke. This action occurs when the paddler extracts the blade quickly from the water; the muscles have finished their work, and the athlete is leaning backward. The arms are dropped to take out the oar.

- The final phase is the recovery, which sets the paddler up for the next stroke. The body moves in the reverse of the way it did for the drive. The hands are pushed away from the body, the body pivots forward from the hips, and the seat is slid into the catch position. As the paddler moves the oar backward through the air, she feathers it by twisting the wrists so that the blade moves parallel to the water. This action lessens wind resistance.

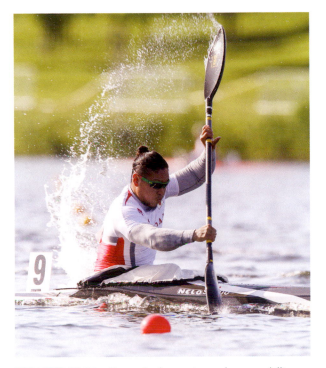

FIGURE 13.12 Frontal plane view when paddling a kayak.

Boris Streubel/Getty Images

The blade needs to be squared up again before the next catch.

AT A GLANCE

Movement Through a Fluid

- To move through a fluid (particularly water) effectively, the two key aspects are to increase propulsion and reduce resistance.

- When kicking a ball, if the impact of the kick is off center, spin can be created on the ball. In the right conditions this spin can then alter the path of the ball (Magnus effect), causing the ball to follow a curved path.

- To reduce the resistance of the craft as it moves through the water, horizontal or vertical movements need to be reduced as much as possible to reduce form drag, wave drag, and to a lesser extent surface drag.

Paddling a Watercraft Characteristics and Mechanical Principles

Technique	Mechanics
To start, the paddler sits in the kayak with the backside as far back as possible in the seat and the knees comfortably bent. If multiple foot wells are available, the athlete straightens the legs all the way out and then brings them back one well. Legs that are too straight may put a strain on the lower back. If the knees are bent too far, the athlete may end up knocking the kneecaps during paddling. Paddlers should work on their strokes in calm water so that they can concentrate on the proper execution of the strokes with the paddle.	A stable and safe ergonomic position is required before paddling the kayak. An anatomically incorrect position can dramatically affect posture and not allow achievement of the most effective force generation. More important, a potentially harmful body position can cause serious injury and damage to the musculoskeletal system.
To find the correct hand placement on the paddle, athletes start with the hands about shoulder-width apart and centered. With the center of the paddle placed on the top of the head, the elbows should form slightly less than a 90° angle. There should be equal amounts of paddle shaft and blade beyond the two hands.	Paddling involves the shoulders and torso more than the arms; with the arms at approximately 90° the paddler is keeping the arms in a relaxed position. Mechanically, the paddler wants to create a lever with the upper body that will generate maximum force to the end of the paddle. When this force is applied to the water, the resultant action is forward propulsion of the watercraft.
Paddlers make the common stroke, the forward power stroke, by leaning forward to put the paddle in the water near the toes. They pull the blade straight back along the boat until it reaches the hip. They then lift the paddle and repeat on the other side.	The basic paddle stroke is a forward power stroke. Placing the paddle blade in the water near the toes reduces the lateral force production and directs the resultant force in the desired direction of travel. The kayaker pulls the paddle blade back alongside the boat to approximately the hip, because after this point the force generated by the anatomical lever rapidly diminishes. To maintain symmetry, the kayaker lifts the paddle blade and takes a stroke on the other side.
Some paddles have the blades offset, sometimes called feathered. Advanced paddlers often use these types of paddles. A different technique must be used to get both blades in the water. If the paddle is a right-hand control (when the right blade is held vertical, the left blade "scoop" is up), the right hand stays tight and the left hand is loose. The way to learn the process is to hold the paddle tight in the right hand and loose in the left. Using the right hand, the paddle blade is rotated back and forth; it should slide through the left hand. The next step is to take a stroke on the right, cock the right wrist back (left hand staying loose and somewhat open), take a stroke on the left, and so forth. If a left-hand control paddle is being used, the process is reversed: The left hand stays tight and the right loose.	A feathered paddle presents less surface area for the wind to catch during the recovery phase of the stroke. There is no other mechanical reason for or advantage to this type of paddle.
The hands should be relaxed during paddling, because the paddle will stay safely in the hand with a medium amount of grip. Sitting posture should be good, the torso should be kept vertical, and a footrest position should be chosen that will allow the knees to be slightly bent. For greater efficiency, not only the arms but also the torso and shoulders should be used. A long paddle will allow a longer stroke; a shorter paddle will give a shorter, faster stroke.	Too strong a grip will cause unnecessary stress within the surrounding muscles, which will deplete the energy system more quickly than necessary. Maintaining correct anatomical posture allows the musculoskeletal system to operate as designed, placing the least amount of stress on the system. Using a longer paddle increases the length of the lever, and if the paddle can be moved with the same amount of linear velocity (or the same stroke rate), a faster boat velocity will result. Alternatively, if a faster stroke rate is desired, then a short paddle lever is required.

REFERENCES

Formosa, D.P., M.G.L. Sayers, and B. Burkett. 2014. "Stroke-Coordination and Symmetry of Elite Back-stroke Swimmers Using a Comparison Between Net Drag Force and Timing Protocols." *Journal of Sports Sciences* 32 (3): 220-28.

Hendricks, S., D. Karpul, and M. Lambert. 2014. "Momentum and Kinetic Energy Before the Tackle in Rugby Union." *Journal of Sports Science and Medicine* 13 (3): 557-63.

Kempton, T., A.C. Sirotic, and A.J. Coutts. 2014. "Between Match Variation in Professional Rugby League Competition." *Journal of Science and Medicine in Sport* 17 (4): 404-407.

Kompf, J., and O. Arandjelović. 2017. "The Sticking Point in the Bench Press, the Squat, and the Deadlift: Similarities and Differences, and Their Significance for Research and Practice." *Sports Medicine* 47 (4): 631-40.

McKean, M.R., and B.J. Burkett. 2014. "The Influence of Upper-Body Strength on Flat-Water Sprint Kayak Performance in Elite Athletes." *International Journal of Sports Physiology and Performance* 9 (4): 707-14.

Tolfrey, V., B. Mason, and B. Burkett. 2012. "The Role of the Velocometer as an Innovative Tool for Paralympic Coaches to Understand Wheelchair Sporting Training and Interventions to Help Optimise Performance." *Sports Technology* 5 (1): 20-28.

www Visit the web resource for review questions and practical activities for the chapter.

APPENDIX

Units of Measurement and Conversions

TABLE A.1   Basic Dimensions and Units in the SI System

Quantity	Symbol	SI unit	Unit abbreviation
Time	t	second	s
Length	l	meter	m
Mass	m	kilogram	kg

TABLE A.2   Derived Quantities and Dimensions Used in Mechanics

Quantity	Symbol	SI unit	Unit abbreviation	SI basic units
Area	A	square meter	m^2	m^2
Volume	V	cubic meter	m^3	m^3
Density	ρ	kilograms per cubic meter	kg/m^3	kg/m^3
Velocity	v	meters per second	m/s	m/s
Acceleration	a	meters per second per second meters per second squared	m/s/s m/s^2	m/s/s m/s^2
Angle	θ	radian	rad	dimensionless
Angular velocity	ω	radians per second	rad/s	1/s
Angular acceleration	α	radians per second per second radians per second squared	rad/s/s rad/s^2	1/s/s $1/s^2$
Momentum	L	kilogram meters per second	kg·m/s	kg·m/s
Force	F	newton	N	$kg·m/s^2$
Weight	W	newton	N	$kg·m/s^2$
Impulse	$F\Delta t$	newton second	N·s	kg·m/s
Pressure	P	pascal	Pa	$kg/m·s^2$
Torque	T	newton meter	Nm	$kg·m^2/s^2$
Moment of inertia	I	kilogram meter squared	$kg·m^2$	$kg·m^2$
Angular momentum	H	kilogram meter squared per second	$kg·m^2/s$	$kg·m^2/s$
Work	U	joule	J	$kg·m^2/s^2$
Energy	E	joule	J	$kg·m^2/s^2$
Power	P	watt	W	$kg·m^2/s^3$

Tables A.3 through A.11 are used to convert from one unit of measure to another for a variety of mechanical quantities. Some instruction on reading these tables is warranted. In the leftmost column of each chart are the units you wish to convert from. The top row of the chart indicates the units you wish to convert to. First, read down the left column and find the unit you wish to convert from. Then, read across the row for that unit until you get to the column for the unit you wish to convert to. Multiply this number by your measure to get the units listed in the top row of that column. For example, in the time conversion chart, to convert 86 min to hours, look in the left column and find the minutes row. Read across to the right until you get to the cell in the hour column. The number in this cell is 0.016667. Multiply 86 min times 0.016667 h/min and get 1.43 h. The conversion factors listed in boldface type are exact conversions.

Mathematically, each cell expresses the following:

1 unit of leftmost column = × units of top-row cell

For our time example,

$$1 \text{ min} = 0.016667 \text{ h}$$

Reading across the whole minute row,

$$1 \text{ min} = 60 \text{ s} = 1 \text{ min} = 0.016667 \text{ h} = 0.00069444 \text{ day}$$

TABLE A.3   Time Conversions

Time	Second	Minute	Hour	Day
1 s =	1	0.016667	0.00027778	0.000011574
1 min =	60	1	0.016667	0.00069444
1 h =	3,600	60	1	0.041667
1 day =	86,400	1,440	24	1

TABLE A.4   Length Conversions

Length	Inch	Foot	Yard	Mile	Centimeter	Meter	Kilometer
1 in. =	1	0.08333333	0.027778	0.000015782	2.54	0.0254	0.0000254
1 ft =	12	1	0.33333	0.00018939	30.48	0.3048	0.0003048
1 yd =	36	3	1	0.00056818	91.44	0.9144	0.0009144
1 mi =	63,360	5,280	1,760	1	160,934.4	1,609.344	1.609344
1 cm =	0.39370	0.032808	0.010094	0.0000062137	1	0.01	0.000010
1 m =	39.3708	3.280840	1.093613	0.00062137	100	1	0.001
1 km =	39,370.8	3,280.840	1,093.613	0.621371	100,000	1,000	1

TABLE A.5   Mass Conversions

Mass	Ounce (mass)	Pound (mass)	Slug	Milligram	Gram	Kilogram
1 oz (mass) =	1	0.0625	0.0019426	28,349.523125	28.349523125	0.028349523125
1 lb (mass) =	16	1	0.031081	453,592.37	453.59237	0.45359237
1 slug =	514.78479	32.17405	1	14,593,903	14,593.903	14.593903
1 mg =	0.000035273	0.0000022046	0.000000068521	1	0.001	0.000001
1 g =	0.035273	0.0022046	0.00068521	1,000	1	0.001
1 kg =	35.27396	2.204623	0.068521	1,000,000	1,000	1

TABLE A.6　Angle Conversions

Angle	Degree	Radian	Revolution
1° =	1	0.017453	0.0027778
1 rad =	57.29578	1	0.15916
1 revolution =	360	6.28319	1

TABLE A.7　Velocity Conversions

Velocity	Feet per second	Miles per hour	Meters per second	Kilometers per hour
1 ft/s =	1	0.68182	0.3048	1.09728
1 mph =	1.46667	1	0.44704	1.60934
1 m/s =	3.28084	2.23694	1	3.6
1 km/h =	0.91134	0.62137	0.27778	1

TABLE A.8　Force Conversions

Force	Pound	Newton
1 lb =	1	4.44822
1 N =	0.22481	1

TABLE A.9　Work or Energy Conversions

Work or energy	Foot-pounds	Joules
1 ft-lb =	1	1.35581
1 J =	0.73756	1

TABLE A.10　Pressure or Stress Conversions

Pressure or stress	Pounds per square inch	Pascal
1 lb per square inch (psi) =	1	6,894.757
1 Pa =	0.000145038	1

TABLE A.11　Torque Conversions

Torque	Inch-pound	Foot-pound	Newton meter
1 in.-lb =	1	0.083333	0.11298
1 ft-lb =	12	1	1.35581
1 Nm =	8.85074	0.73756	1

GLOSSARY

In this glossary we've tried to simplify explanations and terminology as much as possible. To help you understand these mechanical terms, examples are provided that refer to the movement of athletes or sport implements, such as javelins and volleyballs.

acceleration—The rate of change of velocity. An athlete can accelerate, decelerate, or have zero acceleration. In the latter case, the athlete can be either motionless or moving at a uniform rate.

agonist—Muscles whose contractions cause or help action to occur; means *"mover."*

airfoil—The cross-section of a wing. Airfoils come in various shapes. Many have a more curved upper surface than undersurface. Others are symmetrical. They are designed to produce lift.

angle of attack—The angle between the long axis of an object (e.g., a javelin or an airfoil) and the direction of fluid (e.g., air) flowing past. On a discus, the angle of attack is the angle by which the leading edge is raised or lowered relative to the airflow passing by the implement.

angular momentum—For an athlete, angular momentum is determined by the athlete's mass, the distribution of the athlete's mass (how stretched out or compressed the athlete's body mass is relative to the athlete's axis of rotation), and the rate at which the athlete is rotating. In mechanical terms the angular momentum of any object is determined by the object's mass times the distribution of its mass times its angular velocity.

angular motion—Motion that is circular or rotary. Somersaulting, twisting, rolling, swinging, rocking, spiraling, and pirouetting are all forms of angular motion.

angular velocity—The rate of spin of an athlete or object. This term takes into consideration the number of revolutions, the period, and the direction of rotation. An example of angular velocity is 360° in 1 min in a clockwise direction.

antagonist—A muscle whose contraction opposes those muscles that cause movement (i.e., agonists). In sport, an antagonist is an opponent.

apex—The highest point in a trajectory. In a high jump or a dive the apex is the topmost point of the athlete's flight path.

Archimedes' principle—Named after Archimedes, a Greek mathematician. Archimedes' principle states that "the buoyant force acting on an object is equal to the weight of the fluid that the object displaces." If an athlete weighs more than the weight of the water that the athlete takes the place of, then the athlete will sink. Conversely, if an athlete weighs less than the weight of the water that the athlete takes the place of, the athlete will float.

axis of rotation—An imaginary line that passes through the center of rotation of an object or athlete. When an athlete is rotating in the air, the athlete's axis of rotation passes through the athlete's center of gravity. The same occurs for inanimate objects such as volleyballs, baseballs, and javelins. When in contact with the ground, an athlete's axis of rotation and center of gravity are often in different places.

balance—The ability of an athlete to control his movements for a particular purpose.

balance point—The point at which the force of weight is equal on either side.

base of support—The area formed by the outermost points of contact between an object or athlete and the supporting surfaces. The base of support does not necessarily have to be beneath an object.

Bernoulli's principle—Named after Daniel Bernoulli, a Swiss mathematician. Bernoulli's principle states that "pressure exerted by a fluid is inversely proportional to its velocity." Therefore, fluids (e.g., air and water) exert less pressure the faster they move. Conversely, fluids exert more pressure the slower they move.

biomechanics—A science that studies the effects of energy and forces on the motion of living organisms.

boundary layer—The layer of fluid that contacts the surface of an object or athlete when moving in a fluid (e.g., air or water). All objects moving through the air have a boundary layer. Swimmers can have a boundary layer of air passing over those parts of their bodies moving through the air as well as a boundary layer of water for those parts of their bodies moving through the water.

buoyancy—The tendency of a fluid to exert a lifting effect on an object or athlete who is wholly or partially submerged in the fluid. Buoyancy exists in water, air, and other gases.

buoyant force—An upward force (i.e., opposing gravity) exerted on an immersed object by the fluid surrounding it. See also Archimedes' principle.

cadence—The rate per unit time, that is, the number of times that an event will happen during a fixed period.

center of buoyancy—A point at which a buoyant force acts on an immersed object. The center of buoyancy for an athlete is usually higher on the body than the athlete's center of gravity.

center of gravity—A point at which the mass and weight of an object or athlete are balanced in all directions. It is also that point where gravitational forces are centralized. The center of gravity for males is usually higher than that for females.

centrifugal force—When objects or athletes are made to rotate, their inertia constantly tries to make them travel in a straight line. The pull of inertia competes against the inward pull of centripetal force. A common name for the pull of inertia in this situation is centrifugal force. Centrifugal force is really inertia in disguise.

centripetal force—Any rotating object has a force that acts toward its axis of rotation. This inward force is called a centripetal force.

closed skills—Skills performed in a predictable, controlled environment. Synchronized swimming is an example of a closed skill.

conservation of angular momentum—A rotating object or athlete will continue to rotate with constant angular momentum unless an external torque is applied that increases or decreases this angular momentum. During the flight from the tower to the water, a diver will have virtually the same amount of angular momentum from takeoff to contact with the water.

conservation of linear momentum—In any collision between two objects or two athletes, the total amount of momentum before the collision is the same after the collision. Momentum is conserved.

conserve—A term to describe if something is not gained or lost, it is conserved.

deceleration—A decrease in velocity by an athlete or object.

density—Weight per unit volume (or mass per unit volume). The amount of substance contained in a particular space—the greater the amount is, the greater the density is. Muscles and bones are more dense than fat.

discrete skills—Skills that have a defined start and end point; these skills are often nonrepetitive.

dissipation of kinetic energy—In any collision between two objects or two athletes, the kinetic energy brought into the collision is dissipated in the following manner: Some kinetic energy is used by the objects and the athletes in performing work on each other during the collision. But kinetic energy is also lost as heat and noise. See also the law of the conservation of energy.

drag—A force produced by the relative motion of an object or athlete through a fluid (e.g., water or air).

dynamics—A term to describe motion that is undergoing change, that is varying.

eccentric thrust—A force that does not pass through the center of gravity of an object and that tends to cause rotation. Gymnasts, divers, and figure skaters use eccentric thrust to initiate rotation.

elasticity—The ability of an object to recover its original shape after being deformed. Clay is nonelastic. Golf balls, bows used in archery, and modern composite pole-vault poles are all highly elastic.

energy—In mechanics, energy is the ability to do mechanical work. Among the several types of energy are heat, chemical, electrical, and mechanical. The more energy an object or athlete has, the greater the force with which it can shift or deform another object or athlete.

feedback—An evaluation of the performance that is positive and includes suggestions for improvement.

first-class lever—A lever in which the axis is positioned between the force and the resistance. A first-class lever is able to magnify force, balance force with resistance, and magnify speed and distance. First-class levers cause a direction change; the force moves in one direction and the resistance moves in the opposing direction.

flexibility—The range of motion in an athlete's joints. Flexibility is also used to describe the flex of an object like a pole-vault pole.

follow-through—Motions in a skill performance that are performed after the force-producing actions are completed. A follow-through progressively and safely dissipates the momentum and energy generated by an athlete in a skill performance.

force—Any influence that tends to change the state of motion of an object or its dimensions. Force applied by an athlete does not necessarily have to produce movement.

force arm—The perpendicular distance existing between the line of force application and the axis of rotation.

force vector—A symbol indicating the magnitude and direction of an applied force. Force vectors are frequently represented graphically by an arrow. The head of the arrow indicates the direction of the applied force, and the length of the arrow represents the magnitude of the applied force.

force-producing movements—These are the actions used by an athlete to generate force during the performance of a skill.

form drag (also called profile drag, pressure drag, and shape drag)—When an athlete (or an object) travels through a fluid, high pressure occurs in front where the athlete contacts the fluid head on. Low pressure occurs immediately to the rear of the athlete. The greater the difference is between high and low pressure, the greater the form drag is.

friction—A force that acts in opposition to the movement of one surface on another. Friction can be of various types (e.g., static, sliding, and rolling friction).

frontal axis—An axis that runs from the front to the rear. In the human body, the frontal axis passes by way of the center of gravity from the abdomen to the back of the athlete. At the center of gravity, the frontal axis forms right angles where it meets the long axis (from head to feet) and the transverse axis (from hip to hip).

fulcrum—An axis or hinge about which a lever rotates.

gravitational acceleration—The rate of acceleration of an object or athlete toward the earth. Commonly stated as 9.8 m/sec² (32 ft/sec²).

gravitational potential energy—The energy that an object possesses by virtue of being within the earth's gravitational field and above the earth's surface. The more mass an object has and the greater the height it is above the earth's surface, the more gravitational potential energy the object possesses. A diver has more gravitational potential energy when standing on the 10 m tower than when standing on the 3 m board. But a massive diver standing on the 3 m board may possibly have more gravitational potential energy than a lean diver standing on the 10 m tower.

gravity—The force of attraction that moves or tends to move objects with mass toward the center of a celestial body such as the earth or the moon. All objects that have matter have mass and exert a gravitational attraction. The more mass an object has, the greater the attraction is.

ground reaction force—The equal and opposite force exerted in reaction to a force exerted against the earth. An athlete pressing against the earth with a certain force causes the earth to respond both equally and in the opposite direction with a force of the same magnitude.

horsepower—A term introduced by inventor James Watt and used as a measurement of power in the imperial (English) system of measurement. One horsepower is the ability of an object or athlete to move 550 lb through a distance of 1 ft in a period of 1 sec. One horsepower equals 745.7 watts.

human movement—The study of how the human body moves and the influence of movement on the human body.

hydrostatic pressure—The pressure (i.e., force or area) exerted by a fluid (e.g., air or water) to support its own weight or the weight of an object or athlete immersed in the fluid. Atmospheric pressure is 14.7 lb per square inch at sea level. This pressure decreases with altitude. Sea water, which is virtually incompressible, increases in pressure by 14.7 lb per square inch for every 33 ft in depth. To this must be added 14.7 lb per square inch for the weight of the atmosphere resting on the surface of the ocean.

impact—A collision between two or more objects.

impulse—Force multiplied by the time during which the force acts. Athletes vary the amount of muscular force and the period over which this force is applied.

inertia—The tendency of an object or athlete either to stay at rest or to move continuously in a straight line at a uniform velocity. Inertia is directly related to mass. A more massive athlete or object has greater inertia than one with less mass. See also Newton's first law.

key elements—Distinct physical actions that join together to make up a phase of a sport skill.

kinematics—Describes motion; how something is moving.

kinetic energy—The ability of an object or athlete to perform work by virtue of being in motion. Doubling the mass of an object increases its kinetic energy twofold. Doubling the velocity of an object increases its kinetic energy fourfold.

kinetic link (kinetic chain) principle—Simulating a flail-like action or the cracking of a whip. The kinetic link principle is used to accelerate lightweight objects to tremendous velocities in such skills as javelin throwing and baseball pitching. Excellence in throwing requires a progressive increase in velocity from one limb segment to the next, starting at the most massive and ending at the least massive.

kinetics—Describes the forces that create motion or are generated from moving.

laminar flow—A flow pattern in fluids characterized by smooth parallel lines (i.e., like laminations in wood). Laminar flow occurs when fluids pass objects and athletes at very low velocities.

law of the conservation of energy—A law stating that the amount of energy in the universe is constant and cannot be created or destroyed, only changed in form. When an athlete slows down and gives up (kinetic) energy, the athlete's energy is transformed into other forms such as heat, noise, and the movement and distortion of whatever the athlete contacts. The total amount of energy remains unchanged.

lever—A simple machine consisting of a rigid object rotating around an axis or on an axis. In an athlete's body, bones, joints, and muscles work together as lever systems.

lift—The force acting on an object in a fluid that is perpendicular to the fluid flow. Lift is not always upward. It can be made to occur in any direction.

line of gravity—A vertical line drawn from an object or athlete's center of gravity to the earth's surface. A line of gravity is often called a plumb line.

linear and angular kinematic analysis—An analysis of human movement that measures position, displacement, velocity, and acceleration.

linear and angular kinetics—An analysis of human movement that measures forces, inertia, and energy.

linear motion—Motion of an athlete or object in a straight line. If all parts of the athlete or the object move in the same direction at the same speed, this motion is called translation.

linear stability—The resistance of a moving object or athlete to being stopped or having its direction of motion altered. Linear stability is directly related to mass and inertia.

linear velocity—The change in distance divided by the change in time for linear movement, usually expressed as meters per second or feet per second.

longitudinal axis—An imaginary line that runs the length of an object. Also called the long axis. When gymnasts twist, they rotate around their longitudinal axis.

Magnus effect—Named after Gustav Magnus, a German scientist. The Magnus effect is the movement of the trajectory of a spinning object such as a baseball or soccer ball toward the direction of spin. If the front, or leading surface, of a baseball is spinning to the left, the Magnus effect causes the trajectory of the baseball to bend to the left.

Magnus force—A lift force created by the spin of an object such as a baseball, soccer ball, or golf ball. See also Magnus effect.

mass—The amount of matter, or substance, in an object. A massive athlete has a lot of body mass. Mass is also a measure of an object's or athlete's inertia. A massive athlete will have more inertia than an athlete who is less massive.

mechanics—A branch of physics that deals with the effects of energy and forces on the motion of physical objects.

meter—A unit of length based on the metric system. One meter equals 3.281 ft. The 1,500 m race run in the Olympics is 120 yd short of a mile.

metric system—A decimal measuring system with the meter, liter, and gram as the units of length, capacity, and weight (or mass). Used throughout most of the world, although still not adopted for general use in the United States.

momentum—Quantity of motion. The mass of an object multiplied by its velocity. An increase in the mass or the velocity of an athlete increases the athlete's momentum.

negative acceleration—A decrease in velocity of a moving object or athlete.

newton—A measurement used in the metric system and named in honor of Isaac Newton's contributions to science. A newton is a force that when applied to a mass of 1 kg gives it an acceleration of 1 m/sec². One newton equals .2248 lb.

Newton's first law (the law of inertia)—All athletes and objects have mass and therefore have inertia. Their inertia is expressed by a desire to remain at rest. If a force is applied against them to get them moving, their inertia gives them the desire to travel at the same velocity in a straight line. (Forces applied by gravity, friction, and air resistance naturally change this situation.)

Newton's second law (the law of acceleration)—The acceleration of objects or athletes is proportional to the force that acts on them and is inversely proportional to their mass. A more massive athlete accelerates less than an athlete with less mass when the same force is applied to both.

Newton's third law (the law of action and reaction)—When an object or athlete exerts a force on a second object (or athlete), the latter exerts a reaction force on the first that is both equal and opposite in direction. The law of action and reaction applies irrespective of whether the items applying force to each other are athletes or inanimate objects.

nonrepetitive skills (also called discrete skills)—Skills that have a definite beginning and finish and that do not repeat in a cyclic fashion. Examples of nonrepetitive skills include a hip throw in judo and a slap shot in hockey.

observation—To objectively monitor and then assess the movement of the sport skill.

open skills—Skills that are performed in an unpredictable environment. Examples are skills performed against opponents (e.g., wrestling, soccer) and skills performed in changeable environmental conditions (e.g., skiing, yachting, surfing).

phases—A group of movements that stand on their own and which are joined in the performance of a skill. In the long jump, the following are phases: (a) mental preparation and preparatory motions before the approach, (b) the approach, (c) the takeoff, (d) movements performed in flight, and (e) movements performed during the landing.

positive acceleration—An increase in velocity of a moving object or athlete.

power—The rate at which mechanical work is done. (Mechanical work equals force multiplied by the distance that the resistance is moved.) Power is work divided by the time taken to perform the work.

pressure—Force per unit area. Pressure is the ratio of force to the area over which force is applied. The same force applied over a larger area exerts less pressure per unit area.

projectile—Athletes or objects that are propelled in such a way that they follow a pathway (or flight path) resulting from the force of propulsion.

propulsive drag—A form of drag that acts in the same direction that an object or athlete is traveling. In swimming, athletes are able to move their hands and feet in such a way that the drag force generated by the movement of their hands and feet assists in propelling the athlete through the water.

qualitative—A measure that subjectively describes the quality of the sport skill. These terms are more descriptive than objective.

quantitative—A measure that objectively describes the quantity of the sport skills. These terms are typically numeric.

radius of gyration—The distance from the axis of rotation to the center of mass.

rate of spin—A term that quantifies how fast an object or human is rotating, usually measured in terms of degrees per second or radians per second.

real time—A term used that refers to results or measures that will be available as the event is happening or immediately after the completion of the event.

rebound—An action that can result from a sudden impact or collision. The type of rebound depends on the nature of the impact and the elasticity and motion of the colliding objects.

relative motion—A term applied to the relative motion of one or more objects or athletes as they pass one another.

reliability—A term to describe the level of accuracy or dependability of a measure.

repetitive skills (also called continuous skills)—Skills that have a continuous, repetitious, and cyclic nature such as swimming, cycling, and cross-country skiing.

resistance—An oppositional force that tends to retard or oppose motion.

resistance arm—The perpendicular distance existing between the point of application of force applied by a resistance and the axis of rotation.

resistance training—A form of physical exercise where the key purpose is to overload the muscles by providing a form of resistance (usually a weight) that is greater than what is normally experienced.

resultant—A single vector that results from the combination of several vectors. Several forces acting together on an object can produce a single equivalent force that results from their unified action.

resultant force vector—The direction and amount of a force resulting from several forces acting together on an object.

rolling friction—The friction that occurs when a round object such as a wheel rolls against a contacting surface. Rolling friction occurs in gears, axles, transmissions, chains, and engines.

rotary inertia (also called rotary resistance and moment of inertia)—The tendency of objects or athletes to resist rotation initially and then to want to continue rotating after the turning effect of torque is applied and as long as a centripetal force keeps them following a circular pathway. Rotary inertia varies according to mass and the distance that mass is distributed relative to the axis of rotation. A somersaulting athlete has greater rotary inertia in an extended body position than when pulled into a tight tuck.

rotary stability—The resistance of an object or athlete against rotation or being tipped over; if an object or athlete is rotating, rotary stability is the resistance against forces and torque that might disturb its rotation. Rotary stability is directly related to rotary inertia.

scalar measurement—A measurement on a chosen scale of measures; a quantity that gives magnitude but not direction. For example, 32 km/h (20 mph) is a scalar measurement indicating only speed. Velocity indicates both speed and direction.

second-class lever—A lever in which the resistance is positioned between the axis and the force. A second-class lever favors the magnification of force rather than speed and distance. Both force and resistance move in the same direction.

sequence—The order in which the key components of the activity are executed, or the order in which muscles contract to create movement.

skill—A movement pattern designed to satisfy the demands of a sport or specific activity.

skill objectives—The goals or outcomes that occur as a result of the performance of a skill. The rules of a sport determine the skill objectives.

sliding friction (also called kinetic friction)—The friction generated between two surfaces that are sliding (i.e., moving) past each other.

speed—An athlete's or an object's movement per unit of time without any consideration given to direction. For example, 16 km/h (10 mph) is speed with no direction indicated.

sport science—The term to describe the multi-disciplinary field that uses science to understand and subsequently enhance a sporting activity.

spotter—Assistance provided to an athlete during difficult or dangerous phases of a skill. Used frequently in gymnastics.

stability—Resistance to the disturbance of balance. Athletes are able to increase or decrease their stability according to the requirements of the skills they perform.

stance phase—The phase within a gait cycle when a foot is making contact with the ground.

static friction—The friction that exists between the contracting surfaces of two resting objects. Static friction provides the resistive force opposing the initiation of motion between the two objects. Static friction exists between a nonmoving athlete and the supporting surface.

statics—The mechanical term to describe human movement that is in a constant state of motion.

step length—The distance from the location where one foot makes contact with the ground to the location where the opposite foot makes contact with the ground.

strain energy—A form of potential energy stored in an object when the object is distorted or deformed. An archer's bow and a pole-vaulter's pole store strain energy.

strength—The ability of a muscle (or muscles) to exert force without taking into consideration the time that force is applied.

stride frequency—The number of strides per unit time; see also stride frequency.

stride length—The distance from the location where one foot makes contact with the ground to the location where the same foot makes contact next with the ground.

stride rate—The number of strides per unit time; see also stride length.

surface drag (also called skin drag, viscous drag, and skin friction)—Drag that occurs where a fluid (e.g., air or water) contacts the surface of an object or athlete who is moving through the fluid.

swing phase—The phase within a typical gait cycle in which the leg being observed is not making contact with the ground, usually from toe-off to ground-contact.

technique—A commonly used method by which a skill is performed. In high jump, the Fosbury Flop has made the straddle technique obsolete.

temporal gait characteristics—The collective term used to refer to biomechanical measures of gait.

third-class lever—A third-class lever has the force positioned between the axis and the resistance. Third-class levers favor the magnification of speed and distance. Both force and resistance move in the same direction. Third-class levers are the most common lever systems in the human body.

torque—A rotary, turning, or twisting effect produced by a force acting at a distance from the axis of rotation. The initiation of rotation always requires the application of torque.

trajectory—The flight path of an object or athlete.

transfer of angular momentum—A similar term to the conservation of angular momentum. The key difference is the term *transfer*, which highlights how the angular momentum can move from one segment to another.

translation—In mechanics, translation occurs when all parts of an object move the same distance in the same period. This situation is highly unusual because it would mean that absolutely no rotary motion is occurring.

transverse axis—An axis that runs from side to side. In the human body, the transverse axis passes through the center of gravity from one hip to the other. At the center of gravity, the transverse axis forms right angles where it meets the long axis (from head to feet) and the frontal axis (from front to back—abdomen to back).

turbulent flow—A disturbed, nonlaminar flow pattern in a fluid.

uniform acceleration—Acceleration that is regular. Uniform acceleration occurs when velocity increases at a regular rate.

uniform deceleration—A regular or uniform loss of speed or velocity per unit of time. An object or athlete that decelerates by a speed or velocity of 1.5 m/sec (5 ft/sec) for every second that passes would have a uniform deceleration of −1.5 m/sec (−5 ft/sec).

validity—How well (accurate) the measure is compared to what is being measured.

velocity—The speed of an athlete or object in a given direction. Velocity can change when direction changes even though speed may remain uniform.

viscosity—A term describing the stickiness of a fluid and the extent by which a fluid will resist a tendency to flow.

volume—The amount of space occupied by an object or athlete.

vortex—Also called vortices. Vortexes are whirling, swirling masses of fluid. They occur in liquids (e.g., water) and in gasses (e.g., air).

wake—The track left by the rear or trailing edge of an object as it moves through a fluid. A wake is left by ships, swimmers, and airplanes.

wave drag—The drag created by the action of waves at the interface between two fluids. In sport, wave drag occurs where air and water meet.

weight—The force exerted by the earth's attraction on an object or athlete.

windup—Preliminary motions performed by an athlete that increase the distance and the period through which force is applied. A windup is a term commonly used in the description of striking or throwing skills.

work—An expression of mechanical energy. Work is force multiplied by the distance through which an object or athlete moves. A diver climbing the steps of the tower performs work because the diver's mass is raised a certain distance. No mechanical work is performed in isometric exercises because an athlete applies force against an object but causes no movement to occur.

INDEX

Note: The italicized *f* and *t* following page numbers refer to figures and tables, respectively.

ABOUT THE AUTHOR

Brendan Burkett, PhD, is a professor at the University of the Sunshine Coast in Queensland, Australia. He received his undergraduate and master's degrees in engineering and attained his doctorate in biomechanics. Burkett's teaching specializations are biomechanics, sport coaching, and performance enhancement. His research revolves around technology developments in human health and performance, and he has written more than 150 peer-reviewed articles and 180 conference publications for journals in sport science, biomechanics, and coaching.

As an international elite athlete, Burkett represented Australia for 13 years as a swimmer and was the Paralympic champion, world champion, world-record holder, and multiple medalist in the Commonwealth Games and Australian national championships. He served as the Australian team captain for the 1996 Atlanta Paralympic Games and as the flag bearer for the opening ceremonies of the Sydney 2000 Olympic Games. Following his sporting career Burkett has been the sport scientist for the Australian Paralympic team for the 2002, 2006, 2010, 2010, and 2014 World Championships and the 2004, 2008, 2012, and 2016 Paralympic Games. He has attended the past eight consecutive Paralympic Games, from 1988 to 2016.

Burkett has received several awards, including the Australia Day Sporting Award and the Order of Australia Medal (OAM), as well as induction into the Sunshine Coast Sports Hall of Fame, Swimming Queensland Hall of Fame, and the Queensland Sports Hall of Fame.